T0199692

Machinery Adhesives for Locking, Retaining, and Sealing

MECHANICAL ENGINEERING

A Series of Textbooks and Reference Books

EDITORS

L. L. FAULKNER

*Department of Mechanical Engineering
The Ohio State University
Columbus, Ohio*

S. B. MENKES

*Department of Mechanical Engineering
The City College of the
City University of New York
New York, New York*

1. Spring Designer's Handbook, *by Harold Carlson*
2. Computer-Aided Graphics and Design, *by Daniel L. Ryan*
3. Lubrication Fundamentals, *by J. George Wills*
4. Solar Engineering for Domestic Buildings, *by William A. Himmelman*
5. Applied Engineering Mechanics: Statics and Dynamics, *by G. Boothroyd and C. Poli*
6. Centrifugal Pump Clinic, *by Igor J. Karassik*
7. Computer-Aided Kinetics for Machine Design, *by Daniel L. Ryan*
8. Plastics Products Design Handbook, Part A: Materials and Components; Part B: Processes and Design for Processes, *edited by Edward Miller*
9. Turbomachinery: Basic Theory and Applications, *by Earl Logan, Jr.*
10. Vibrations of Shells and Plates, *by Werner Soedel*
11. Flat and Corrugated Diaphragm Design Handbook, *by Mario Di Giovanni*
12. Practical Stress Analysis in Engineering Design, *by Alexander Blake*
13. An Introduction to the Design and Behavior of Bolted Joints, *by John H. Bickford*
14. Optimal Engineering Design: Principles and Applications, *by James N. Siddall*
15. Spring Manufacturing Handbook, *by Harold Carlson*
16. Industrial Noise Control: Fundamentals and Applications, *edited by Lewis H. Bell*
17. Gears and Their Vibration: A Basic Approach to Understanding Gear Noise, *by J. Derek Smith*
18. Chains for Power Transmission and Material Handling: Design and Applications Handbook, *by the American Chain Association*

OTHER VOLUMES IN PREPARATION

Machinery Adhesives for Locking, Retaining, and Sealing

GIRARD S. HAVILAND

Loctite Corporation
Newington, Connecticut

CRC Press

Taylor & Francis Group
Boca Raton London New York

CRC Press is an imprint of the
Taylor & Francis Group, an **informa** business

First published 1986 by Marcel Dekker, Inc.

Published 2019 by CRC Press
Taylor & Francis Group
6000 Broken Sound Parkway NW, Suite 300
Boca Raton, FL 33487-2742

© 1986 by Taylor & Francis Group, LLC
CRC Press is an imprint of Taylor & Francis Group, an Informa business

First issued in paperback 2019

No claim to original U.S. Government works

ISBN 13: 978-0-367-45165-3 (pbk)
ISBN 13: 978-0-8247-7467-7 (hbk)

Visit the Taylor & Francis Web site at
http://www.taylorandfrancis.com

and the CRC Press Web site at
http://www.crcpress.com

Library of Congress Cataloging-in-Publication Data

Haviland, Girard S., [date]
    Machinery adhesives for locking, retaining, and sealing.

    (Mechanical engineering ; 44)
    Includes bibliographies and index.
    1. Joints (Engineering) 2. Adhesives. 3. Machine
parts. 4. Metal bonding. I. Title. II. Title:
Machinery adhesives. III. Series.
TJ1320.H38 1986   621.8'6   85-25314
ISBN 0-8247-7467-1

*Dedicated to the conquest of inner space and the
improvement of productivity in the world,
and to the Loctite Corporation and its
customers who are working to
make this happen.*

Preface

The spaces left after the assembly of threaded, flanged, and press-fitted parts have always been a source of trouble. The amount of metal-to-metal contact between threads and heavy press fits varies between 20 and 30% of the total area involved, which means that 70 to 80% is do-nothing space. Shifting threads, moving flanges, and fretting cylindrical fits will loosen, leak, and allow parts to fail catastrophically. Before 1956, the ills of inner space were treated mechanically by eliminating as much space as possible with closer fits and finer finishes, at ever-increasing cost and with rapidly diminishing results. Jamming the space with an intermediate material such as caulk or solder was effective but awkward and sometimes, with solder, thermally destructive and irreversible. Since 1956 the Loctite Corporation has specialized in organic materials that cure exclusively in these airless spaces, adding convenience to function and resulting in benefits of cost and reliability that modern engineering cannot ignore. The Loctite Corporation's assistance in preparing this book has been critical to its accuracy and completion.

The object of this book is to guide the designer, process engineer, or mechanic in selecting and using anaerobic machinery adhesives effectively. It is my hope that students of engineering also will benefit from this book, so I will include the "why" as well as "what" and "how" about these materials. By understanding machinery adhesive technology, it is possible to be innovative in its application. The early innovators were called "Loctite Charlies" because of their addiction for adhesive solutions to mechanical fitting problems. I am indebted to them for showing the way and providing examples that I have categorized into generic classes. These generic examples are shown in the appropriate chapters and in the design hints section, in order to start the inventive juices flowing.

Most adhesive applications require a systems approach to be successful. Although machinery adhesives have been formulated to achieve success in ordinary industrial environments, it is still necessary to consider surfaces, application method, cure system, and test methods before the final benefits are realized. Professor Gerald Schneberger of the GMI Engineering and Management Institute, an astute teacher in the adhesive and coating field, had a tongue-in-cheek way to point out the needs of a systems approach for adhesives. The following are his "Seven Sins of Commission" (or "How to Hate Anything to Do with Adhesives"):*

1. Skip the test in your facility. Just use the average results off of the vendor's data sheet. "Typical" properties show that you can squeak by; and your application is typical.
2. Keep the old joint design even if it doesn't suit an adhesive. Somewhere there must be *something* with high peel and high tensile strength too.
3. Assume that on-line temperature, humidity, and cleanliness will cure the adhesive the same as the conditions in your laboratory.
4. Expect your adhesive to be stronger than steel and just as strong in water as it was in your prototype test.
5. Keep the vendor out of your shop. After all, his experience is gained from people like you, and your process is so unique it must be kept secret.
6. Surprise your workers at the last minute to impress them with how progressive you were to have the process worked out without their knowing.
7. Keep the design, materials, and production people at arm's length. They aren't the best of buddies anyway, and they have a habit of asking tough questions. Besides, if they get interested in your project, they will just delay your getting things going.

What I will do is give you, the designer, engineer, user, and assembler, enough data and knowledge about the products to plan your systems approach. When your requirements are not covered by the data in the book, it should be evident that you need assistance from specialty chemical formulators. They can match a material to your requirements or give you appropriate data.

This is a book of "why," "how," and "what" for standard machinery adhesives. A lot of the data is empirical, derived scientifically in

---

*Adapted with permission.

the laboratory. Where theoretical and analytical approaches help in understanding and designing, they are given. Data are assembled in tables and graphs for ready reference.

As with any book of technology, this one builds on what was done by others. I am deeply indebted to those who worked with me and came before me. I especially want to thank the Loctite Corporation, originator of this technology, and all the men and women who share its history. (For more about these people, I highly recommend the book *Drop by Drop: The Loctite Story*, by Ellsworth Grant, Loctite Corporation, Newington, Conn., 1983.)

My gratitude and thanks for the assistance of writing talents greater than mine go to Bruce Burnham of Bruce Burnham & Associates for spearheading the illustration production, and to Gale Sorensen of Barbeau Associates for proofing and correcting script.

Girard S. Haviland

Contents

Machinery Adhesives for Locking, Retaining, and Sealing

Chapter 1

General Information About Machinery Adhesives

I. INTRODUCTION

Ever since machinery has been built, designers and machinists have faced the problem of fitting assemblies to minimize the inner space that allows leaking and moving, or working, of apparently tight parts. Inner space is either the clearance that exists between parts to allow for their assembly, such as in a threaded assembly, or the space that cannot be filled because a press fit can produce only peak-to-peak contact of surface irregularities, thus leaving a substantial 70 to 80% of noncontact inner space (Fig. 1.1).

In the early 1800s special machines were developed to turn the bores and parts of cannon and steam cylinders which, in James Watt's original experiments, had "close" fits of about 0.06 in. To produce tight joints, strips of leather, hemp, and clay were used to caulk the joints. Since then, machinery to fit cylindrical parts has required more and more precision at higher and higher costs (Fig. 1.2).

Designers, engineers, and machinists can now eliminate this problem on stationary fits by using machinery adhesives. Typical stationary fits would be a bolt in a threaded hole, a press fit of a ball bearing on a shaft, or a clamped flange sealing internal machine parts. In contrast to structural adhesives, which are used as the primary holding means in a structure (often in direct tension), machinery adhesives are generally used in rigid cylindrical assemblies in a shear or compressive mode to eliminate leakage and provide a noncreeping joint.

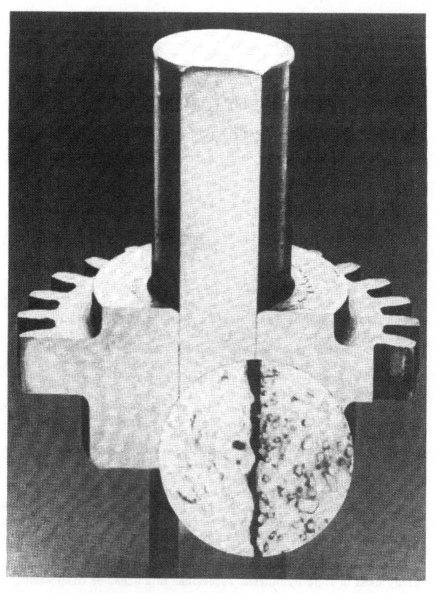

(a)

FIGURE 1.1   Inner space of a heavy press fit and a class 2 thread.

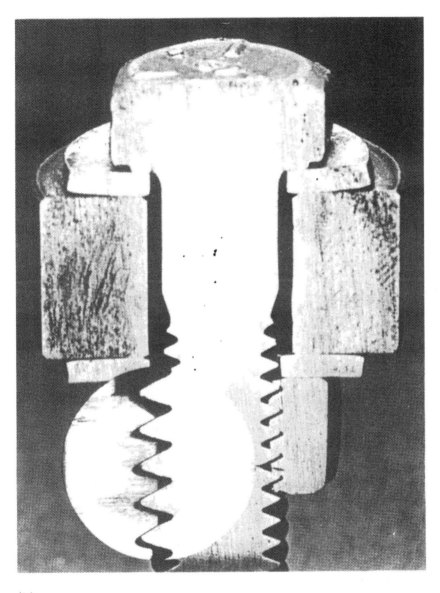

(b)

FIGURE 1.1 (Continued)

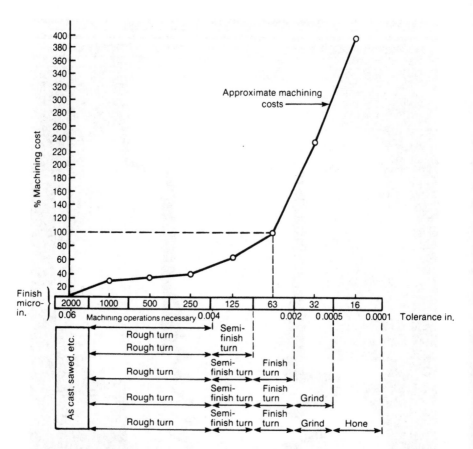

FIGURE 1.2   Relative machining costs, surface finishes, and tolerances. (Adapted from *Tool and Manufacturing Engineers Handbook*, 3rd edition, Society of Manufacturing Engineers, Detroit, Michigan, 1976.)

1.1   Machinery Adhesive Subgroups and the First-Choice Materials

Thread lockers for straight and tapered threads
     Grade M, N, and O, or preapplied MM, NN, SS, and TT
Stud setters
     Grade K, L, or O
Friction improvers for press fits
     Grade S, T, or U
Retaining compounds for slip fits
     Grade S, T, or U

Thread sealants for tapered and straight threads
    Grade W (tapered thread), N, or preapplied NN (straight)
Sealants and friction improvers for flat surfaces
    Grade X or Y
Porosity sealants for castings, welds, and powder metals
    Grade R or S
Shims for flat or cylindrical surfaces
    Grade Z

2. WHAT THEY ARE LIKE

Brazing, soldering, and adhesives all look the same to the designer however much they differ in detail. Separate parts are joined by an intermediate material—a liquid different from the base materials being joined—which is drawn by wetting and capillary action into the joint, where it hardens to form the bond. Soldering and brazing materials were the first materials used for cylindrical fitting.

In soldering and brazing, the hardening is a physical change of state. In adhesion the hardening is usually the result of a chemical change that transforms a liquid into a dense, hard polymer. The result is the same: a bonded, sealed joint where inner space is filled to achieve the result. Although solders and brazes are still used extensively because of their strength and liquid fill characteristics, they require temperatures of 600 to 1200°F to melt the solder to wet the part, which can be a serious drawback. Some metals cannot be wetted at all, and at these temperatures many parts are distorted or destroyed.

In 1964 a new line of free-radical-curing materials that were stabilized by the presence of air was introduced to the machinery market and aptly given the name *anaerobic.** These compounds, although complex in their chemistry, are simple to use and durable after cure. For the most part, anaerobic materials cure at room temperature within minutes after they are confined between parts such as a bolt and nut, a bearing and bore, or two flanges, or in pipe threads. Cure is initiated by a free radical of iron or copper on the surface.

Modern chemistry and engineering have refined the formulations so that many parameters of organic machinery adhesive can be controlled, including viscosity, lubricity, cure speed, shear strength, modulus, ultimate strength, impact resistance, and chemical resistance, as well as minor characteristics such as color and fluorescence. The original anaerobic free-radical-cure system has been improved in some cases with moisture cures, ultraviolet cures, and two-component, dry-to-the-touch preapplied films. The preapplication method involves the micro-encapsulation of the resin or the activator, which is mixed into a

---

*They were developed by the American Sealants Company, founded in 1954, which is now the Loctite Corporation.

slurry. The slurry is applied to parts away from the assembly line and allowed to dry before assembly takes place. The material releases a fast-curing liquid during the assembly of threads in a manner analogous to squeezing water from a sponge. Preapplied machinery adhesives are Type VI Grades MM, NN, SS, and TT.

3. WHERE THEY ARE USED

1. Thread locking and sealing were the first uses for these anaerobic materials. Filling the threads with a hard, dense material prevents self-loosening of nuts and bolts as in Fig. 1.3, which illustrates a nut and bolt secured with an anaerobic resin.

2. Thread sealing and preventing of corrosion are other purposes of thread locking and sealing materials. The disassembled badly rusted bolt in Fig. 1.4 shows how corrosion was prevented in the threads by a Grade N adhesive.

3. Press fits and bore close-in on bushing mountings can be avoided by the use of a machinery adhesive (Fig. 1.5). Slip fit bushings may be aligned from the shaft rather than the bore.

FIGURE 1.3 Inner space of threads secured with cured anaerobic resin.

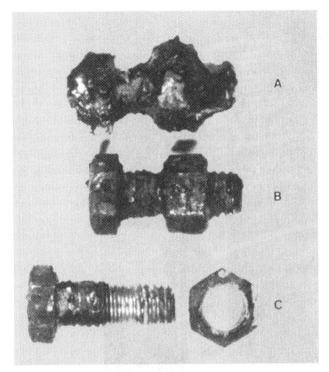

FIGURE 1.4 Threads sealed on a corroded nut and bolt: (A) as corroded for six months in salt spray, (B) after wire brushing, (C) disassembled, showing clean threads protected by machinery adhesive N.

4. Ball bearing assembly using machinery adhesive allows accurate alignment from shafts with relaxed tolerances and fits of the bores. The assembly can be done by hand without presses or hammers. No distortion of the bearing or housing takes place (Fig. 1.6).

5. Key and keyway fitting prevents fretting and loosening from reversal torques which occur on most driven shafts (Fig. 1.7).

6. Most economical is the retaining and sealing of cylindrical parts such as cup plugs, shafts, rotors, gears, pulleys, and oil seals that may have been previously press fitted with mediocre results or expensive failures (Fig. 1.8). Fragile shafts can be assembled and secured without bending or misalignment.

7. Rigid shafts can be assembled easily without powered equipment (Fig. 1.9).

8. Impregnating powder metal parts, porous castings, and welds prevents underplating corrosion, holds pressure, and assists in machining by preventing tool wear (Fig. 1.10). Many parts that needed plating for appearance or protection could not be made from

FIGURE 1.5   Slip fitted bushing.

FIGURE 1.6   Adhesively mounted miniature dental turbine drill
bearings (cutaway) illustrate how alignment from the shaft can allow
operation at 400,000 RPM without shake or loosening.

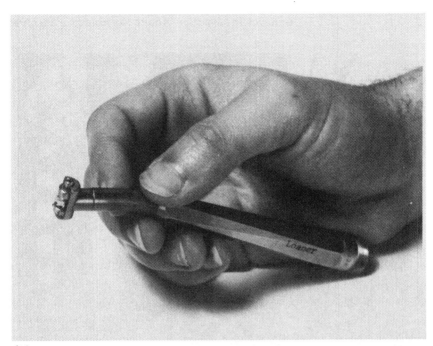

(a)

(b)

FIGURE 1.7   Keys are fitted with adhesive on a crown gear of a heavy-duty drive.

any process that allowed absorption of plating solutions. Impregnation and sealing with anaerobics is an excellent way to overcome this deficiency and take advantage of the economies of such net shape processes as pressed powder metal.

4. HOW THEY ARE SELECTED

Like plastics and metals that are molded and heat-treated by the manufacturing purchaser, adhesive products must be applied and cured by the user. They become part of his processing operation. Therefore, the user must consider not only the adhesive's cured properties but also the fluid properties and the variables of application and cure. In this respect adhesives are akin to paints.

There are three material properties that must be considered in selecting proper machinery adhesives:

1. Flow properties of the liquid
2. Conditions and characteristic of the cure
3. Properties of the cured material

(a)

(b)

FIGURE 1.8   Cup or core plug sealing and retaining on an automotive head (a) using a rotospray applicator (b).

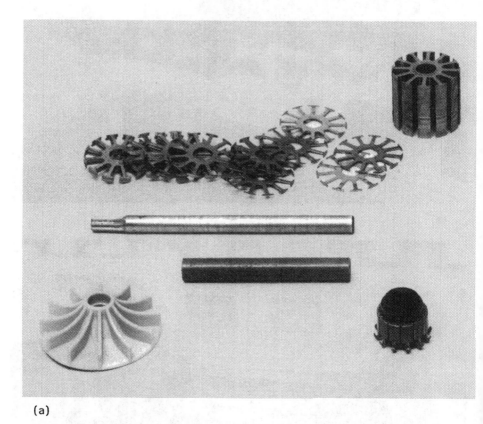

(a)

FIGURE 1.9 Bonded with machinery adhesives are the rotor, shaft, armature, fan, and commutator on a fractional-horsepower double-insulated electric motor.

(b)

FIGURE 1.9 (Continued)

FIGURE 1.10   Plated parts, untreated (above), and impregnated with
an anaerobic (below), both after the standard plating process.

4.1   Liquid Properties

The rheology or flow of the liquid must be considered for the type of
joint being filled, the method of filling, and the processing conditions.
A very thin—1 to 100 centipoise (cP)—material can be wicked into a
heavy press fit provided the material doesn't start to thicken and
cure before it completely penetrates. For hand application before assembly, a viscosity of 1000 to 10,000 cP might be appropriate (a thin
to thick syrup). If machine coating of a bore is desired, then a thixotropic material is used. (Thixotropy is the flow-rate-sensitive characteristic that makes ketchup stay in the bottle until it begins to flow,
then suddenly gush after flow starts.) Often these materials are disc-
sprayed onto a bore. It is possible for a fine-sprayed ring of thixotropic material to stay on the inside of a bore for many minutes or
until a plug is pushed into place (Fig. 1.11).
    Processing conditions may also affect the consideration of liquid
properties. These conditions may dictate the use of a dry material.

FIGURE 1.11   Rotospray (right) application to a cup plug bore (left).

For instance, on an automobile final assembly line, where upholstery and finish paints are present, the use of liquids may be discouraged or forbidden. In this case, thread locking materials selected are almost always Type VI, preapplied dry to the touch (Table 1.1).

4.2   Cure Conditions and Speed

Different products develop strength at different speeds, depending on the surface condition of the joined parts, the gap between parts, the pressure and mixing during assembly, and the temperature and humidity.

A typical cure speed is given in Table 2.1 to help in judging relative speeds. In many cases, a slow-curing product is selected to allow for assembly of complex parts. Fast cures are selected for high production rates.

When selecting speed of cure for high production, be sure to measure the speed on actual parts under production conditions and allow a safety factor of two for variability. In other words, a 10-second

TABLE 1.1   Properties of the Liquid

Thread-treating materials are produced in three liquid types as listed in MIL S-46163.

Type I: Sealing, standard viscosity—Newtonian[a]

    Grade K   500 cP or mPa.s   3/8-1 in. 10—24 mm threads

    Grade L 7000 cP or mPa.s   5/8-1 in. 16—24 mm threads

Type II: Lubricating—thixotropic[a]

    Grade M 5000 cP or mPa.s   #2-1/2 in. 2—12 mm threads

    Grade N 5000 cP or mPa.s   1/4-3/4 in. 6--20 mm threads

    Grade O 7000 cP or mPa.s   3/8-1 in. 10—24 mm threads

Type III: Wicking, thread, and porosity sealing[a]

    Grade R 15 cP or mPa.s   #2-1/2 in. 2—12 mm threads

Other useful machinery adhesives have been developed since the 1974 Mil. Spec. 46163

Type IV: Newtonian, high strength or high temperature[a]

    Grade S (high strength)   100 cP or mPa.s   #2-1/2 in. 2—12 mm

    Grade T (high temperature) 7000 cP or mPa.s   1/4-3 in. 6—75 mm

    Grade U (high strength)   2000 cP or mPa.s   1/4-3 in. 6—75 mm

Type V: Pastes for tapered thread and flat flange sealing

    Grade W   550,000 cP or mPa.s   1/4-1 1/4 in. 9—32 mm pipe

    Grade X 3,800,000 cP or mPa.s   0.010 in., 0.25 mm max. gap

    Grade Y   850,000 cP or mPa.s   0.010 in., 0.25 mm max. gap

    Grade Z 1,200,000 cP or mPa.s   0.010 in., 0.25 mm max. gap

Type VI: Preapplied, dry-to-the-touch thread lockers[b]

    Grade MM   (low strength, silver)

    Grade NN   (medium strength, green)

    Grade SS   (high strength, red)

    Grade TT   (high temperature, yellow)

TABLE 1.1   (Continued)

Primers or activators for use with Types I-V

Primer N   Increases cure speed and doubles gap cured on the
             thinner materials.

Primer T   Increases cure speed. See Chap. 2 for the effect of
             accelerators on curing and cured properties.

Lubrication

Where precise clamp loads are required, the lubricating Type II
materials are preferred. They may be used over a lightly oiled
surface. All materials have predictable friction factors which
should be considered for good clamp load control (T = KDF,
Chap. 2).

---

[a]The use of accelerators and primers doubles the gap-curing capability of Types I—IV.
[b]Type VI materials are usually applied by a bolt-converting manufacturer and shipped as part of the bolt. Upon assembly the dry
sponge material exudes a quick curing liquid, which turns solid and
provides the same bolt security and sealing as the normally applied
liquids. Activator is included in the dry film so no external activator
or primer need be applied.

average fixture time, as determined experimentally, should not be tied
to a machine that allows less than 20 seconds before parts are unclamped (Fig. 1.12).

Primer or activator N or T is used to give active surface results on
inactive parts. For further curing data see Chap. 2, Sec. 3.

4.3   Cured Properties

Strength and environmental resistance should be considered together.
For instance, if a material is to be used near the upper limit of its
temperature rating, its strength may be as little as 50% of the room
temperature rating shown in the tables. In such a case, a higher-
strength material should be selected. Strengths also vary according
to the surfaces being bonded. The thread locking selection charts,
Figs. 2.2 and 2.3, should be consulted for particular surfaces.
Medium strength (800 psi, 5.5 MPa) is usually sufficient for most

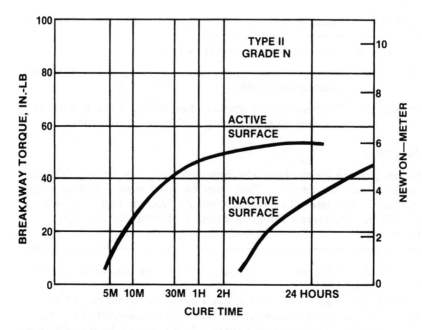

FIGURE 1.12   Typical cure speed curve—Type II, Grade N.

thread locking and cylindrical fits. Higher strengths are used only
for severe duty and larger, high-strength parts that can be disassembled without damage.

Other cured properties may be important, depending on the application. Check those that are necessary and select a material accordingly.
You can tell the approximate strength of the liquid formulations by
their color. Low strength is purple or medium blue, and high is red
or green.

4.4   Uniformity of Results

Machinery adhesives are difficult to test in the compressive mode in
which they are often used. Precision compression testing requires
thick sections of materials, which are difficult to produce with anaerobics because of the stability of liquids in large sections. Large chunks
of force-cured materials are liable to be highly stressed from uneven
shrinkage; therefore, their relative strengths are usually tested in
shear. High-modulus materials in shear give a fairly large variation in
results. Shear failures can be either cohesive (the most consistent) or
adhesive, that is, they break free of the surface of the test specimen.

Shear type results will depend on surface adhesion with all the ram-
ifications of material, finish, cleanliness, and gap. Predictability from
published data will be on the order of magnitude of ±25%. The prudent
engineer will test production parts to achieve more precise results.
This not only helps in design but is required where quality control
departments must check whether proper fill and cure have taken place.
In most cases where machinery adhesives are used, a nominal 800 to
1000 psi stress (5.5 to 7 MPa) is sufficient evidence that the compressive load can be supported. In thin films of 0.005 in. or 0.13 mm the
materials will support loads of 35,000 to 55,000 psi (240 to 380 MPa),
which encompasses the tensile and compressive yield strength of low
carbon steel.

4.5 Designing with Published Data

Sheer stress data are produced on pins and collars 0.5 in. or 12 mm in
diameter and on nuts and bolts 3/8" ×16" or M10. When the parts being designed deviate from those conditions of the test specimens by a
substantial amount, the data must be altered to reflect the predicted
conditions. Figs. 1.13 and 1.14 help do that.

Most of the materials are formulated for diametral clearances of
0.003 in. (0.08 mm) or less. Those that perform well over this value
should be downrated unless the data are obtained at enlarged gap.

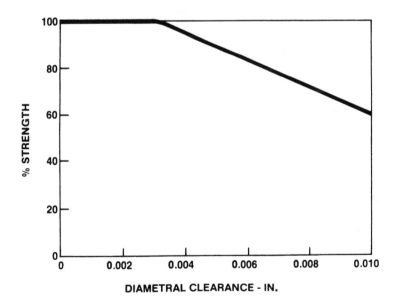

FIGURE 1.13   Strength vs. gap.

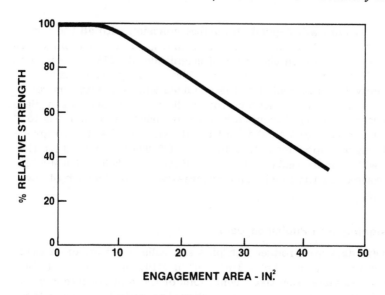

FIGURE 1.14   Shear strength vs. engaged area.

The reason for the decrease in larger gaps is not known, but it is a phenomenon that can be accurately measured and documented (Fig. 1.13).

The decrease in strength when dealing with areas over 10 in.^{2} (65 cm^2) is conjectured to be the result of greater clearance because of poorer geometry of fit, chance of poorer fill because of the greater area involved, and elongation of the parts with greater length of engagement. An extreme example of what minimal engagement does to the average stress is demonstrated by the simple test used for structural adhesives, the lap shear test.* This test uses two pieces of substrate 0.062 in. (1.6 mm) thick by 4" × 1" (100 × 40 mm) (Fig. 1.15).

At first one might conclude that doubling the length of overlap would double the strength of the joint; however, because of stress concentration, doubling the overlap from 0.5 to 1 in. (13 to 25 mm) increases the strength by only 1.65 on steel laps (Fig. 1.16). The reason can be visualized by observing that the stress distribution is entirely different for the doubled-lap length. The peak stress that

---

*American Society for Testing and Materials (ASTM) Method D1002.

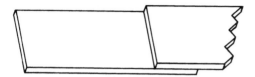

FIGURE 1.15   American Society for Testing and Materials (ASTM)
D1002 lap shear specimen.

initiates failure is the same, but the average stress is less because of
the inability of the lap centers to support any load.

This same phenomenon takes place in cylindrical and threaded fits
as the parts get larger and the length of fit longer. In threads and
cylindrical fits the effect does not become significant until engagements of four times the diameter are encountered.

Within the limits of close fits and good machinery practice, the
rougher the finish the better the apparent adhesion or, more accurately, the mechanical keying (Fig. 1.17). Keying is not effective if
the stress works parallel to the lay of the finish (for further discussion see Chap. 6, Sec. 3.2.1).

There is a definite relationship between screw size and achieved
strength. Figure 1.18 shows that a material rated at 1000 psi (7 MPa)
on 3/8 in. bolts will develop a shear stress on a #2 screw of 1600 psi
(11 MPa). The thread locking performance charts (Figs. 2.2 and 2.3
in Chap. 2) have already taken this effect into consideration. The
reason for the phenomenon is the same as for cylindrical fits: gap and
size effect. A 1½ in. bolt, for instance, will develop only 60% of the
3/8 in. shear rating (Fig. 1.18).

The curve shown in Fig. 1.18 is for a medium adhesion Type II
Grade N, which usually breaks on a nut or bolt in the adhesive,
rather than cohesive, mode. Surface material can give variability of
about two to one.

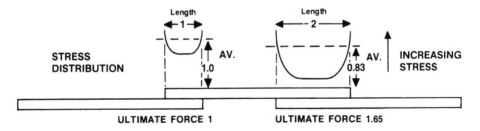

FIGURE 1.16   Stress distribution in loaded laps.

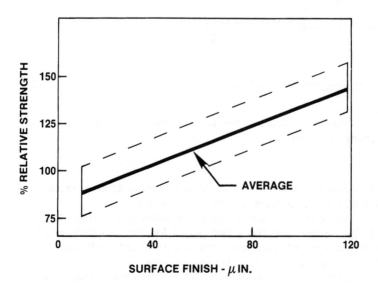

FIGURE 1.17   Strength vs. surface finish.

Grade N is rated on different surfaces in Fig. 1.19. Dry zinc phosphate, which is used extensively for paint adhesion, is a good base for machinery adhesives and gives the highest shear stress values of all the common bolt surfaces.

Although emphasis is understandably on ultimate strength as a rating characteristic, it is essential to consider the fatigue character-

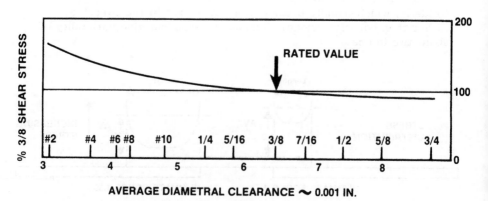

FIGURE 1.18   Strength vs. screw size.

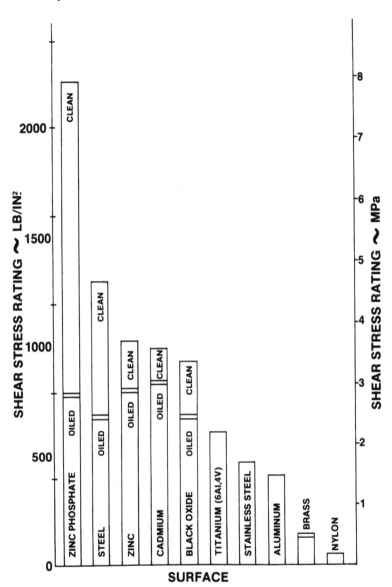

FIGURE 1.19 Strength vs. surface material.

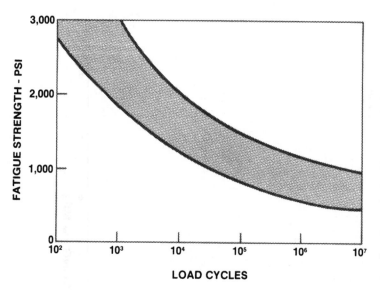

FIGURE 1.20 Fatigue strength versus load cycles.

istic of the materials that will be 10 to 15% of the ultimate, as in
Fig. 1.20. Obviously, from our example of lap shear stress distribution, exacting tests should be conducted on actual parts if service life
involves many cycles of stress; otherwise 10% of ultimate should be the
maximum safe calculated load. Chap. 6 has a further discussion of
fatigue of motor shafts (Fig. 1.20).

Another useful characteristic is the prevailing torque, which on a
nut and bolt is measured as the torque at one-half turn from where
break occurred.* This prevailing torque used to be called locking
torque for thread jamming systems, which existed before anaerobics
were invented. Thread jammers had no break torque. Prevail can be
useful for preventing loss of the fastener if the break has been lost
from retorquing after relaxation of a gasket or from adjusting an adjusting screw. Grade R is formulated specifically to wick into adjusting screws and give high prevailing torque is they are moved. Resistance is substantially lost if parts are completely disengaged but
will maintain itself through dozens of small adjustments.

*The exception is in Mil-S-22473 and 46163 where prevailing torque is
the average of four readings taken one-quarter, one-half, three-
quarters, and one full turn from break. The one-half turn data are
usually within 10% of the average method data.

5. PACKAGING AND HANDLING

5.1 Packaging and Aging

All of the machinery adhesives are anaerobic, which means they are stabilized by oxygen in the air. To maintain stability, bottles and tubes of material are usually not completely filled, leaving air surface over the liquid. The package is made of low-density polyethylene, which allows air molecules to pass through the wall from the outside.

Shelf life is 1 year or more for unopened liquid and paste products and 4 years for the preapplied products (see Table 2.1 for specific times). Within these ages they should meet all specifications for viscosity and cure speed. Although this may sound brief for hardware markets, the useful life of each of these products is usually more because all are fail-safe. Liquid products will gradually slow down in their cure speed but still produce specified cured properties. If cure speed is critical, check the date code on the bottom of the bottle with your supplier. Try some parts. The preapplied products gradually harden on the bolt and become difficult to assemble, like a prevailing torque locker, and thus are easy to inspect. Paste products may harden in the tube center; however, paste that is left will still perform adequately. Elevated temperatures, humidity, and sunlight can accelerate aging, but under normal conditions (60°F or −40°F and 40% RH) the products last many years. They are not harmed by subfreezing temperatures but most become thick and inactive below 40°F so they should be applied at comfortable, shirtsleeve temperatures. Accelerators are generally used for subfreezing cures (see Chap. 3, Sec. 4).

Mixing from one container to another is not recommended, since this action may introduce contamination which can cause set-up or loss of cure speed. If material must be removed from its original container for application, limited amounts should be removed so it is not necessary to return used material to the original container.

Often people ask if different materials can be mixed to achieve a viscosity or color more suitable to a particular application. Although this has been done successfully in a few instances, the successes are more the exception than the rule. At least no violent reactions have been reported. These compounds are carefully formulated combinations of 15 to 20 ingredients, some effective in parts per million. They should not be mixed. For particular requirements of viscosity, color, speed, and strength some manufacturers will formulate specials. These formulators should be consulted.

Refrigeration does not impair the shelf life of anaerobic machinery adhesives, nor is it particularly helpful. If the adhesives would be subject to high temperatures, then it could be beneficial.

5.2   Safety

Even in today's tightly regulated business and safety climate, machinery adhesives are generally considered very safe since they are used
in relatively small quantities and are not aggressively unsafe. Some
of them do contain adhesion promoters that have a low level of toxicity
on ingestion, inhalation, skin absorption, or eye contact. They should
be handled using good industrial hygiene. Since there is some variability from manufacturer to manufacturer, the material safety data
sheet (MSDS) should be consulted before materials are put on line.
For further discussion of safety, refer to Chap. 4, Sec. 6.

5.3   Shipping

Adhesives

No special precaution need be taken for shipping any of the machinery adhesives, either as a liquid or a solid. The Code of Federal
Regulations (CFR) Title 49, Subtitle B definition of "combustible
liquid" is one that has a closed cup flash point of 100°F (38°C) or
less. Any product containing methacrylic acid or any acid base may
be a "corrosive liquid." Neither of these restrictions is pertinent to
the primary suppliers of these materials as all show no restrictions by
the Department of Transportation (DOT) or International Air Trans-
port Association (IATA). To be sure, ask your supplier for the
MSDS.

Primers or Activators

Primers N and T are now classified as "poisonous," "flammable,"
and "combustible" because of their major constituents, 1,1,1-tri-
chloroethane (both N and T) and isopropyl alcohol (T only). These
materials react violently with acetone, sodium hydroxide, or oxygen.
Trichlor and alcohol are common industrial solvents. Trichlor is often
used in vapor degreasers and general cleaning solvents. Isopropyl
alcohol (rubbing alcohol) is used instead of trichlor for cleaning on
parts sensitive to chlorinated solvents and is often used as an anti-
freeze and a cleaner in windshield washer solutions.

Because of the toxic nature of trichlor [toxicity level value (TLV)
350] and the flammability of isopropyl alcohol, the official descriptions
for shipping are as follows:

Primer N shipping regulations
    Type or class
        DOT ORM-A ("other restricted material" Group A)
        IATA Class 6.1 poisonous substance (effective January 1, 1983)

Proper shipping name
  DOT 1,1,1-trichloroethane
  IATA poisonous liquid, n.o.s. (no other specification) (contains
  1,1,1-trichloroethane)

Primer T shipping regulations
  Type or class
    DOT combustible liquid
    IATA Class 3 flammable liquid (as of January 1, 1983)
Proper shipping name
    DOT combustible liquid, n.o.s.
    IATA flammable liquid, toxic, n.o.s. (1,1,1-trichlorethane)

5.4 Metal Corrosivity and Plastic Compatibility

Types I, II, and III are required to pass a corrosivity test according
to Mil-S-46163. Under high-humidity conditions (40% and up), a discoloration can occur that appears to be an oxide but has been dis-
covered to be a discolored film of cured material. The other types may
or may not have adhesion promoters which can affect the surface of
some metals. As soon as cure takes place all effect on the surface
ceases. Thus no tests have ever shown corrosion to be a problem
even when residual stress or adverse environments have been present.
To the contrary, field and laboratory tests in corrosive atmospheres
have shown that the resin-filled inner space has eliminated corrosion
because the crevice or reservoir for holding electrolyte is gone, as
shown in Fig. 1.4.

Compatibility with Plastics

Without testing, plastic compatibility is predictable only in extreme
cases. ASTM Standard Practice D3929 for evaluating the stress cracking of plastics by adhesives using the bent-beam method is used for
evaluating and predicting the effect of adhesives on various materials
at different stress levels. All materials under appropriately high
stress become sensitive to liquid and gaseous intrusion into the grain
boundaries and between molecules. The degree to which this happens
is dependent on the material, the stress, and the environment. Hydrogen embrittlement of steel can occur at low stresses in a zinc or cad-
mium plating bath. A hardened piece of carbon steel (e.g., a high-
performance valve spring in an internal combustion engine) can become embrittled by water at stresses over 100,000 lb/in.^{2} (690 MPa).
Plastics, which have an order of magnitude lower strength, exhibit
similar behavior when under stress even if they are highly crosslinked thermoset materials. Machinery adhesives should not be used

TABLE 1.2   Plastic Compatibility[a]

Compatible	Stress-sensitive; try carefully	Not compatible
Acetals	Polyvinyl chloride (PVC)	ABS
Alkyd		Acrylic
Allyl	Polyurethane	Cellulosics
Amino resins		Polycarbonate
Epoxy		Polyphenelyne oxide (PPO)
Fluroplastics		
Nylon		Polystyrene
Phenolic		Polysulfone
Polyimide/polyamide-imide		Styrene acrylonitrile (SAN)
Polyethylene		
Polyphenylene sulfide		
Polypropylene		
Thermoset polyester		

[a]Trade names are shown in Table 1.3.

on or near any thermoplastic material unless full cure can be assured without touching the plastic. Remembering that the stress level in the plastic is critical to its susceptibility, Table 1.2 can be used for general guidelines only.

Applied stresses can often be avoided until cure and cleanup take place; however, molding stresses from thermal gradients often exceed 1000 psi and, when they do, stress cracking can occur within seconds or minutes of contact with liquid materials. Application should be carefully tried on parts that can be spared. If no cracking or crazing occurs within 24 hours usually the combination is safe. Table 1.3 gives the trade names for most common generic plastics.

TABLE 1.3   Trade Names

ABS (acrylontrile, butadiene, styrene)	Abson—Abtec Chemical Co.
	Cycolac—Borg-Warner Chemicals
	Lustran—Monsanto
	Kralastic—Uniroyal
Acetals	Delrin (acetal homopolymer)—du Pont
	Celcon (acetal copolymer)—Celanese
Acrylic	Acrylite—Cy/Ro Industries
	Lucite—du Pont
	Plexiglas—Rohm and Haas
	Corian—du Pont
Alkyd	Plascon—Allied Chemical
	Glaskyd—American Cyanamid
Allyl plastics	CR 39 (allyl diglycol carbonate)—PPG
	Dapex—Acme Resin Company
	Dapon (diallyl phthalate)—FMC Corporation
Amino resins (urea and melamine formaldehide)	Plaskon—Allied Chemical
	Cymel—American Cyanamid
Celluosics	Tenite—Eastman Kodak
Epoxy	Plascon—Allied Chemical
	Polyset—Morton Chemical Company
	Epon—Shell
	Araldite—Ciba-Geigy
Fluoroplastic	Teflon—du Pont
	Halon—Allied Chemical
	Tefzel—du Pont
	Kynar—Penwalt Corporation

TABLE 1.3   (Continued)

Ionomer	Surlyn—du Pont
Nylon	Zytel—du Pont
	Vydyne—Monsanto
	Capran—Allied Chemical
Phenolic	Plascon—Allied Chemical
	Genal—General Electric
	Durez—Hooker Chemical
	Plenco—Plastics Engineering
PPO	Noryl—General Electric
Polyimide/polyamide-imide	Torlon—Amoco
	Vespel—du Pont
	Kinel—Rhone-Poulenc
Polyethylene	Dylan—Arco
	Alathon—du Pont
	Tenite—Eastman Kodak
	Norchem—Northern Petrochemical
	Marlex—Phillips Petroleum
	Bakelite—Union Carbide
	Microthene—U.S. Industrial Chemicals
	Petrothene—U.S. Industrial Chemicals
Polypropylene	Tenite—Eastman Kodak
	Pro-Fax—Hercules
	Marlex—Phillips Petroleum
Thermoset polyester	Haysite—Haysite Reinforced Plastics Co.
	Rosite—Rostone Corporation

TABLE 1.3 (Continued)

Thermoplastic polyester (PBT, PET)	Versel—Allied Chemical
	Ekkcel—Carborundum
	Tenite—Eastman Kodak
	Celanex—Celanese
	Valox—General Electric
Polycarbonate	Lexan—General Electric
	Merlon—Mobay Chemical
Polypheneylene sulfide	Ryton—Phillips Petroleum
Polystyrene	Dylark—Arco
	Styron—Dow Chemical
	Lustrex—Monsanto
	Bakelite—Union Carbide
Polyvinyl chloride (PVC)	Dacovin—Diamond Shamrock
	FPC—Firestone
	Geon—B.F. Goodrich
Polyurethane	Estane—B.F. Goodrich
	Texin—Mobay Chemical
	Orthane—Ohio Rubber
	Vibrathane—Uniroyal
	Pellethane—Upjohn
Polysulfone	Udel—Union Carbide
Styrene acrylonitrile (SAN)	Tyril—Dow Chemical
	Lustran—Monsanto

5.5   Removal

Like the problem of the fellow who invented the perfect solvent—he
didn't know what to hold it in—the removal of misapplied machinery
adhesives can be difficult. One of the purposes of this book is to
describe how to apply the technology so that parts can be disassembled when necessary. But we all have had the dilemma when ordinary
tools will not remove adhesive. What can be done? Or, if the parts
are disassembled, how can they be cleaned up for reassembly? The
following tips are second best to planning ahead for possible disassembly, but they will be useful to know.

1. Heat—most machinery adhesives will weaken considerably between 400 and 600°F (200—300°C). (Check the rating and go 150°F
over it.) The part, of course, must be able to withstand the
temperature.

2. Impact or cleavage—the materials are generally weakest in
these two modes. Tapping a bearing out of a bore usually gets better
results than does a steady push. Likewise, peeling the corner of a
gasketed cover is more likely to unzip it than impact or pull will.
Once the parts separate, physical brushing or scraping will often complete the job because brittle materials will powder. It is not necessary
to remove all the material for reassembly; the material that is well ad-
hered can be left. New material will be entirely compatible, although
the covered surface will be inert for curing and it is advisable to consider activation for rapid cures.

3. Chemical methods—a methylene chloride (methyl chloroform)
soak will usually release the adhesive and soften the resin. Some
parts (such as assembled bearings and fasteners) will have to be
soaked for up to 24 hours because so little bondline is exposed. Methylene chloride is available in handy spray form from some adhesive
manufacturers.

These are formulations especially made to lift old gaskets, paint,
and adhesives from surfaces. As with any chlorinated solvent, care
should be taken with their use. Follow the directions on the can and
if in doubt try a limited area first. If they will eat a machinery adhesive, they certainly will destroy any thermoplastic or elastomer.

4. Full-strength Lestoil* will remove some exposed machinery adhesives after a 24-hour soak. Most affected are Grades W, X, and Y.

5. Lestoil in a 50% water solution will remove films of preapplied
threadlockers, Grades MM, NN, SS, and TT. Cold solutions will take
overnight. Hot solutions at 150°F (65°C) will take a couple of hours.

6. Oakite[†] stripper 157 (viscous) or 156 (liquid dip) will remove
exposed machinery adhesive after 24 hours of contact or soak.

---

*Noxell Co., Household Products Division, Baltimore, MD.
[†]Oakite Products Inc., Valley Road, Berkeley Heights, NJ 07922.

Stripper #157 is a viscous solvent-alkaline product designed to re-move resistant finishes from aluminum, magnesium, and steel surfaces too large for tank immersion. It meets the requirements of Mil-R-25134A (USAF) "Remover, Paint and Lacquer Solvent Type." It is somewhat reactive on zinc, brass, and copper.

Stripper #156 is a nonchlorinated, di-phase solvent formulation designed for stripping tough synthetic finishes. It is satisfactory for steel and aluminum. It mildly affects the surface of brass and copper but attacks magnesium and zinc.

Both materials are used at full strength and are aggressively alka-line. Manufacturers' recommendations for use and safety precautions should be followed rigorously.

7. Removal of liquid material—liquid or uncured material can be removed with most shop solvents. Chlorinated solvents such as tri-chloroethane or methylene chloride are very effective whether used cold or in a vapor degreaser. Methyl ethyl ketone (MEK), acetone, methyl alcohol, and freon are effective. Water is not. Good safety and hygiene practices should be followed with any of these materials. This includes avoiding skin contact.

So what about skin cleanup? Continuous contact should be avoided, but when contact occurs a dry or waterless mechanic's hand soap ef-fectively releases the material so that ordinary soap and water can remove the residue.

For materials spilled on clothes, normal dry cleaning usually takes care of the problem. A soak in vegetable oil before dry cleaning often helps to remove pigments and dyes.

6. GOVERNMENT AND INDEPENDENT LABORATORIES REQUIREMENTS

6.1 Underwriters Laboratories

In some markets it is desirable and sometimes mandatory that inde-pendent laboratory testing be done to establish the suitability of a material for its intended use. The Underwriters Laboratory (UL)* was founded in 1894 as an independent, nonprofit organization. Its purpose was to evaluate materials, devices, products, equipment, con-struction methods, and systems with respect to hazards affecting life and property.

Testing emphasis is on product and public safety through functional evaluation and follow-up testing by experimental engineers under con-tract to manufacturers, government agencies, and others.

---

*33 Pfingsten Road, Northbrook, IL 60062.

Service and Product Coverage

1. Service categories available.
    a. Product listing service—the UL maintains lists of qualified products and controls the use of its symbol.
    b. Classification service—laboratory evaluation classifies products with respect to specific hazards, limitations, or performance conditions.
    c. Component recognition service—this deals with the evaluation of component parts and materials (such as rubber or plastic) that will later be used in a complete product or system.
    d. Certificate service—certain types of products, such as building materials (e.g., shingles) can't practically bear the UL label and accompanying certificates are provided.
    e. Inspection service—UL's trained inspectors worldwide check out products periodically on a contract basis.
    f. Fact finding and research—projects are conducted by the UL on a contract basis for manufacturers, trade associations, and government agencies.
2. Product groups—lists of products by category are published in directories if in conformance with requirements.
    a. Building materials.
    b. Fire protection equipment.
    c. Fire-resistant material.
    d. Recognized components.
    e. Electrical appliance and utilization equipment.
    f. Electrical construction materials.
    g. Hazardous location electrical equipment.
    h. Marine products.
    i. Classified products (machinery adhesives).
    j. Accident, automotive, and burglary protection equipment.
    k. Gas and oil equipment.

Machinery adhesives Grades K, W, and X are covered under product group 2.i above, which classifies them on a fire hazard basis. The products so covered are marked on the package; the label for Grade K shown in Fig. 1.21 is typical.

The fire hazard class was established for the uncured material according to Table 1.4, which shows classes of familiar materials. For requirements of test methods refer to Bulletin UL 340.

Canadian and British UL facilities are cooperating but independent ventures. Approvals must be sought from each separately.

"Adhesive/Sealant 271," Fire Hazard is small. No flash point in liquid state. Ignition temperature 304°C (579°F). For use in devices handling gasoline petroleum oils, natural gas (pressure not over 300 psig), butane and propane.

FIGURE 1.21 Typical package marking showing the Underwriters Laboratory mark.

TABLE 1.4 Fire Hazard Classification Scale

Numerical fire hazard rating	General classification	Flammability temperature limit °F (°C)
100	With diethyl ether	−49 (−45)
90−100	With gasoline	13 to −48 (−10.6 to −44.4)
80−90	Between ethyl alcohol and gasoline	38 to 14 (3.3 to −10)
70−80	Between ethyl alcohol and gasoline	51 to 39 (10.6 to 3.9)
60−70	With ethyl alcohol	67 to 52 (19.4 to 11.1)
50−60	Between kerosene and ethyl alcohol	83 to 68 (28.3 to 20.0)
40−50	Between kerosene and ethyl alcohol	99 to 84 (37.2 to 28.9)
30−40	With kerosene	129 to 100 (53.9 to 37.8)
20−30	Between paraffin oil and kerosene	256 to 130 (124.4 to 54.4)
10−20	With paraffin oil	440 to 257 (226.7 to 125)
0−10	Less hazardous than paraffin oil	−
0	With water	Noncombustible

Extent of Approval

Approvals do not extend to the products assembled with these materials. For instance, the approval given for a pipe sealant is for the sealant in the uncured state and does not include a pipe assembly. Separate approvals must be sought for the assembled product. For instance, a gas meter assembled with Grade W needs separate approval even though this sealant is approved for such an application.

6.2   U.S. Department of Agriculture

The following products are recognized by the U.S. Department of Agriculture (USDA)* as "chemically acceptable for the use in slaughtering, processing, transporting or storage areas in incidental contact with meat or poultry food products prepared under Federal inspection": Grades N, K, U, and W, and Primers N and T. ". . . Acceptance is valid as long as the corporation and use remain as described to us [USDA ed.] and provided the inspector approves the performance. No endorsement of the material of any concomitant claims is intended."

The products are not edible in liquid form although they become inert after curing. Any usage in food handling equipment requires full cure and removal of excess liquid before sterilization or use of the equipment. Machinery adhesives, as normally used to secure and seal threads, bearings, and press fits, logically qualify as minimal contact elements and are routinely used with local inspector approval for assembly and repair of food-related equipment.

6.3   Food and Drug Administration

The Food and Drug Administration (FDA) approval or listing of products requires that they be either edible or safe in direct contact with food products. Machinery adhesives are not in either of these categories and are therefore neither approved nor disapproved by the FDA. They are not eligible for FDA action.

6.4   National Sanitation Foundation

The National Sanitation Foundation (NSF)[†] does not allow materials in potable water systems or food processors to have extractable chemicals or detectable taste or odor. Machinery adhesives are detectable after cure. However, individual mechanisms using machinery adhesives can be approved after appropriate cleaning and testing.

---

*Senior Staff Officer, U.S.D.A., Compounds and Packaging Section, Chemistry Division—Science, Washington, D.C. 20250.
[†]NSF Building, Ann Arbor, MI 48105.

6.5 Military and Government Specifications

Army and Navy

Government philosophy is changing regarding the issuance and maintenance of the thousands of internally generated specifications used to purchase materials for its own use. The Office of Management and Budget (OMB) of the U.S. Government has issued Circular A-119, which says that the government should stay out of the direct generation of standards. It includes the following directives:

Use voluntary standards in the interests of greater economy and efficiency.

Give preference to voluntary standards over nonmandatory government standards. Review in 5 years all government standards to cancel those that can be replaced by voluntary standards.

Have knowledgeable government employees take part in standards-producing bodies at government expense as authorized agency representatives with the objective of eliminating government specifications.

Allow technical support such as cooperative testing and participation of government employees in the policy-making processes of voluntary standards bodies, including preparation, coordination, and review of standards themselves.

In spite of this directive, new military specifications continue to be used and revised, so it is well to understand the basic procedure of issuance and revision as illustrated by the following example.

1. A government contractor wants to use a new product or material in the design or production of his assembly and needs a specification to cover the properties. He writes for approval from his government contracting office and relays his request to the controlling branch of the U.S. Army, Navy, etc.
2. The agency recognizes the need after enough inquiries have been reviewed and writes or alters a specification. Qualified suppliers are usually contacted for advice in specification preparation.
3. A specification or amendment is issued. Suppliers may certify to it as required.

Note that the supplier does not initiate the action but only assists after action has been started. Users of nonspecified materials can get specification recognition by contacting the Chief of Specifications and Technical Data Board, Code DRXMR-LS Department of the Army, Army Materials and Mechanics Research Center, Watertown, MA 02172.

Air Force and Aerospace

In 1976 a contract was made between Battelle Columbus Laboratories and the U.S. Air Force for the purpose of the preparation and maintenance of Mil-Std-1515 "Fastener Systems for Aerospace Applications." This was in response to the awareness that 20 to 30% of the cost of a primary aircraft structure can be attributed to the procurement and installation of mechanical fastening systems.

In January 1982 a task group was formed to address the use of anaerobic sealing and locking compounds so that they might be included in Mil-Std-1515. Draft of Std 1515 requirement 114 is being circulated for approval per the written procedures of the Aero Mechanical Fasteners Requirements Group (AMFRG). According to the draft, Type I Preapplied and Type II Liquids may be used for the following purposes:

1. For locking and sealing permanently installed fasteners. They should not be used on any threaded device that is normally disassembled during routine maintenance.
2. As a supplement to a primary locking device in single-point structural connections that may be subject to rotation during normal service.
3. As a supplement to primary locking devices in any single point primary structural connection, the loss of which might endanger personnel or the serviceability of the flight vehicle.

Current Military Specifications

Since specification documents are written for specific formulations all "equivalent" formulations may not match every element of the specification. This is especially true of formulations from nonoriginal manufacturers who are attempting to qualify. Some formulations may be a little faster, slower, thicker, thinner, etc. Some deviation in curing properties may be tolerated or even desired if production processes are considered and exceptions are fully documented. For instance, Specification Mil-S-22473D was written to cover the original Loctite Corporation letter grade materials. Mil-S-46163 covered materials in an improved form that had faster, oil-tolerant cures. Some are thixotropic and include lubricating properties. Where the earlier specification was in use for many years before Mil-S-46163, many drawings and user's documents were committed to the older materials. The Grumman Corporation has conducted temperature and surface compatibility tests to determine the functional interchangeability of letter grades (22473) and the number grades (46163) as shown in Table 1.5.

TABLE 1.5   Military Specification Interchange

Mil-S-22473D grades	Approximate equivalents	
	Mil-S-46163 grade	Mil-R-46082A type
AA	R	
A	R	
D	O, L, or K	
AV[a]	O, L, or K	
AVV	L	
B	O	
C[a]	N	
CV	N	
CVV	N	
E	M	
EV	M	
H[a]	M	
HV	M	
HVV	M	
JV	None	
	S	I
	None	II
	None	III

[a]Grumman Corporation compared AV with O, C, with N, and H with M. Other equivalent products have strength and viscosity differences that make pretesting advisable.

Although some grades and types have been omitted in this handbook, the Grades K through TT carry on the lettering system of Mil-S-46163 and cover most requirements of viscosity, cure speed, and strength. Special requirements may necessitate contact with a supplier who has formulating and engineering capability.

6.6   Nuclear Use and Requirements

Effect of Irradiation on Strength

Irradiation tests of Grades K, L, M, N, R, S, T, U, V, W, X, and Y have been done with gamma radiation for total doses up to 200 megarads. The effect on strength as measured by break/prevail torques on steel nuts and bolts was in general rather minimal. Usually, the break was the same to 40% lower whereas the prevailing torque was 100 to 170% of the nonirradiated parts. This would imply that some embrittlement had taken place in a manner similar to heat aging.*

Sulfur and Chlorine Content

To avoid the creation of corrosive elements and steel embrittlement during irradiation, the Nuclear Regulatory Commission requires that organic compounds contain no more than 200 parts per million (ppm) of chlorine or 1500 ppm sulfur (Atomic Energy Commission Regulatory Guide 1.37 or ANSI N 45.2.1). To achieve these levels special processing must be followed. None of the materials in this guide should be used in nuclear situations unless specifically certified to meet the NRC requirements. Those materials so certified may be used in the secondary side of a generating plant, that is, the electrical generating side. On the primary side, none may be used in the containment area but certified materials may be used in the rad-waste control system, instrumentation, controls, etc., under high radiation (200 megarads) where temperatures are limited to 150°C (300°F). Consult your supplier for nuclear-rated materials.

---

*Isomedix Inc., 25 Eastman Road, Parsippany, NJ 07054, and Loctite Corp., 705 North Mountain Road, Newington, CT 06111 (Report T-1221).

Chapter 2
Engineering Data Bank

1. SUMMARY OF PROPERTIES

The properties listed in Table 2.1 are typical only of the particular
materials tested, which were made by the Loctite Corporation. In most
cases they are averages of many specimens from many batches. Similar materials are made by other organizations. The values in the tables
are useful for design purposes where normal safety factors are used
and prototype tests can confirm results. They should not be used for
receiving or engineering specifications.

The headquarters of manufacturers listed in industrial catalogs are
listed below. Local directories should be consulted for the closest service.

Apple Adhesives, Inc.
8000 Cooper Avenue
Ridgewood, NY 11385
(513) 332-3533

Bostik Division, Emhart
Middleton, MA 01949
(617) 777-0100

Devcon Corporation
Danvers, MA 01923
(617) 777-1100

Fel Pro Chemical Products Division
7450 North McCormick Boulevard
Skokie, Illinois 60076
(312) 761-4500

Henkel KGaA
P.O.Box 1100
D-4000 Dusseldorf 1, West Germany
0211/7971

Hernon Inc.
37-23 27th Street
Long Island City, NY 11101
(212) 784-8001

Loctite Corporation
705 North Mountain Road
Newington, CT 06111
(203) 278-1280

Rocol Limited
Rocol House, Swillington
Leeds LS268BS, England
Garforth (09738) 2261

Master Bond
P.O.Box 522
Teaneck, NJ 07666
(201) 343-8983

Three Bond Co., Ltd.
1456 Hazama-cho, Hachioji-shi
Tokyo, Japan
0426 (61) 1337

Permabond International
480 South Dean Street
Englewood, NJ 07631
(201) 567-9494

2. THREAD LOCKING MATERIAL SELECTION

The selection of material is based almost entirely on the removal torque
and not the severity of duty. The reason for this is the emphasis on
torque as the one convenient way to inspect for presence and cure of
the material. Additionally, it has been proven that a line-to-line fit
with very low (almost zero) break torque has given very good resistance to self-loosening. Any low-shrink, high-modulus, cured liquid
will give good performance regardless of its lack of adhesion to the
threads. Anaerobics, of all curing materials, do this most conveniently.
Our selection will be based on making the break-loose torque as close
to the tightening torque as possible. In that way the removal tools and
effort will imitate the tightening process.

2.1   Selection to Match Proof Strength of Fastener

Tightening Torque

Tables 2.2 through 2.6 provide information to calculate the allowable loads and tightening torques for various surfaces and materials.
The formula $T = KDF$ is explained in Chap. 5, Sec. 1.1., or approximate tightening torques can be read from Fig. 5.2.

TABLE 2.1 Properties of Machinery Adhesives— Summary

	ASTM or other spec.	UNITS	TYPE I GRADE K 271*	TYPE I GRADE L 277	TYPE II GRADE M 222	TYPE II GRADE N 242	TYPE II GRADE O 262	TYPE III GRADE R 290	TYPE IV GRADE S 609	TYPE IV GRADE T 620	TYPE IV GRADE U 680
1.1 LIQUID PROPERTIES											
Rheology			Newtonian	Newtonian	Thixotropic	Thixotropic	Newtonian	+Wicking+	Newtonian	Thixotropic	Newtonian
Chemical Designation			Anaerobic Dimethacrylate	Anaerobic Dimethacrylate	Anaerobic Methacrylate	Anaerobic Methacrylate	Anaerobic Methacrylate	Anaerobic Methacrylate	Anaerobic Dimethacrylate	Anaerobic Dimethacrylate	Anaerobic Dimethacrylate
Viscosity	D2556	mPa.s	750	7000	1000	1100	1500	15	100	7000	1250
Suggested clearance		inch							0 to 0.003	0 to 0.004	0.001–0.003
Bolt Range		inch	3/8-1	5/8-1+	#2-1/2	1/4-3/4	3/8-1	#2-1/2	#2-1/2	1/2-3	1/4-3/4
Max Gap Cure		in.(mm)	.025(.6)+T	.025(.6)+T	.022(.6)	.022(.6)	.025(.06)	.016(.4)	.003(.08)	.015(.4)	.015(.4)
Color			Red	Red	Purple	Blue	Red	Green	Green	Green	Green
Specific Gravity			1.12	1.12	1.05	1.05	1.1	1.07	1.10	1.15	1.08
Flash Point	Cleve.OC	°Fm(°C)	Above 200 (93)		Above 200 (93)	Above 200 (93)	Above 200 (93)		Above 200°F (93°C)		
Shelf Life (70°F±20°F 22±11°C)		Months	24	24	24	24	24	24	24	24	24
Corrosivity	Mil-S-22473		None	None	None	None	None	None	None	None	None
Lubricity K Factor on Oily Phos			0.12	0.14	0.14	0.11	0.10	0.16	0.26	0.17	0.18
See also TABLE 2.6											
1.2 CURING			Oily steel Nuts and Bolts			Oily steel Nuts and Bolts			Steel Pins and Collars		
Fixture		Min.	15	30	15	10	3	2	20	85	—
20% of Ult.		Min.	30	60	30	15	10	3	24	24	25
Full cure		Hrs.	24	24	24	24	72	24		72	72
Primer for inactive surface or gap cure			T or N	T or N	None or T	None or T	None or T	T or N	N or T	None+Heat, T	N or T
Primed 20% ult.		Min.	2	7	6	5	22	8 or 4	7	30	1 or 10
1.3 CURED PHYSICALS	SPEC.		GRADE K	GRADE L	GRADE M	GRADE N	GRADE O	GRADE R	GRADE S	GRADE T	GRADE U
Thermal Cond.	General 0.14	BTU/hr.ft²°F									
Thermal Conductivity	General 0.13 est.	W/m°C									
Coeff. Therm. Exp.		10⁻⁵/°C	5–10						5–10	10	10
1.4 MECHANICAL PROPERTIES											
Mod. of Elast. Ten.	D412		General 300,000 Lb/in² (2100 MPa)						General 300,000 Lb/in² (2100 MPa)		
Shear Strength, St. Pin/Collar		Lb/in²							3000	3500	4500
		MPa							21	24	31
Shear Strength, Nut & Bolt		Lb/in²	2900/360	2100/2100	500/120	1300/270	1200/2400	880/2800	3000/—		2000/—
—Break/Prevail—St. Degreased		MPa	20/2.5	14/14	3.4/0.8	9/1.9	8.3/17	5.5/1.9	21/—		14/—
Shear Strength, Nut & Bolt —Break/Prevail-Zn. Phos.		Lb/in²							3000/—		2000/—
Compressive Str.Thin Film		1000Lb/in²	125		125	125			175		
		MPa	860		860	860			1200		

*Commercial number, Loctite Corp.

TABLE 2.1 (Continued)

ASTM or other spec.	UNITS	TYPE I GRADE K 271*	TYPE I GRADE L 277	TYPE II GRADE M 222	TYPE II GRADE N 242	TYPE II GRADE O 262	TYPE III GRADE R 290	TYPE IV GRADE S 609	TYPE IV GRADE T 620	TYPE IV GRADE U 680
1.5 ELECTRICAL PROP.										
Dielec. Str. — D149	V/mil	General 250 →						General 250 →		
1.6 HEAT/COLD RESISTANCE	UNITS									
General Rating	°F(°C)	300°F (149°C) →						300°F(149°C)	450°F (204°C—300°F) →	
Hot Str.@ Rated Temp. — 72°F	% of	75	50	30	30	85	90	30	68	50
Heat Aged at Rated Temp.	1000 Hours	→						1000 Hours →		
Cold Strength-72°F(22°C) Air	% Ret.	30	50	70	20	30	90	125	91	120
Minus 100°F(-73°C)-Acetone+CO₂	% Ret.	100	100	100	100	100	100	100	100	100
Minus 320°F(-196°C)—Liq.N₂	% Ref.	75	135	170	130	87	90	—	—	—
1.7 CHEMICAL RESISTANCE 188°F (87°C)		1000 Hours →						30 Days at 188°F (87°C) →		
Air Reference — Mil-S-22473D	%	100	100	100	100	100	100	100	100	100
Motor Oil	%	70	83	67	100	100	86	100	60	110**
Water	%	110	64	35	27	100	74	40	94	63
Glycol/Water (50/50%)	%	65	59	27	30	98	74	50	—	86
Transmission Fluid	%	96	90	88	100	100	90	100	87	140
Gasolene	%	50	90	57	95	86	90	70	—	110
Skydrol	%	70	90	82	95	78	90	100	—	—
Gasohol	%	65	—	—	87	—	100	—	—	—
Trichlorethane	%	—	—	—	—	—	—	—	71	110
Butyl Alcohol	%	—	—	—	—	—	—	—	85	110
Phosphate Ester	%	—	—	—	—	—	—	—	100	110
Toluene	%	—	—	—	—	—	—	—	58	100
Isopropyl Alcohol	%	—	—	—	—	—	—	—	—	—
1.8 MILITARY SPECIFICATION		Mil-S-46163 →					Mil-S-46163 →		—	—

* 140% at 300°F

** 140% at 300°F

TABLE 2.1 (Continued)

ASTM or other spec.	UNITS	TYPE V GRADE W	TYPE V GRADE X	TYPE V GRADE Y	TYPE V GRADE Z	TYPE VI GRADE MM	TYPE VI GRADE NN	TYPE VI GRADE SS	TYPE VI GRADE TT
Commercial Number (Loctite Corp.)		567	518	510	660	203	202	204	201
1.1 LIQUID PROPERTIES									
Rheology		Paste	Gel	Gel	Paste	——Dry Preapplied——		——Dry Preapplied——	
Chemical Designation		——Anaerobic Dimethacrylate——				——Anaerobic Dimethacrylate——			
Viscosity—D2556	mPa.s	550,000	3,800,000	850,000	1,200,000	——Solid Sponge——		——Solid Sponge——	
Suggested clearance	inch	0—.005	0.01 Max.	0.01 Max.	0.001–0.020	[Class 2&3A&B up to 5 td/i		[Class 2&3A&B up to 5td/in	
Bolt Range	inch	1/2–1 Pipe	1/8–1/2 Pipe			[Metric 6H and 3Pitch		[Metric 6H and 3Pitch	
Max Gap Cure	in.(mm)	0.02(0.5)	0.01(0.25)	0.02(0.5)	0.02(0.5)				
Color		White	Purple	Red	Dark Grey	Silver	Green	Red	Yellow
Specific Gravity		1.16	1.07	1.1	1.1	1.1	1.1	1.1	1.1
Flash Point—Cleve.OC	°Fm(°C)	Above 200 (93)			Above 200°F	(93°C)		Above 200°F (93°C)	
Recertification Time	Months	20	24	24	20(<50 ml)	48	48	48	48
Corrosivity-Mil-S-2M1-S-22473		None	None	None	None	None	None	None	None
Lubricity K Coeff. on Oily Phos. See TABLE 2.6		0.08	0.18	0.19	0.18	0.11	0.13	0.18	0.15
1.2 CURING		——St. Pipe Th'ds Flat Laps——			Pin & Collar	——Oily steel Nut & Bolt——		——Oily Steel Nut and Bolt——	
Fixture	Min.	---	---	---	10	10	10	10	10
20% of Ult.	Min.	240	240	1200	10	24	24	24	30
Full cure	Hrs.	24	48	Stage A 90 / B 72@200°F	12	72	45	72	72
Primer for inactive Surface or Gap Cure	Min.	N	N	N	N or 250°F	[Cure is independant of surface.]		[Cure is independent of surface.]	
Primed 20% ult.	Min.	5	5	15	1 or 30				
1.3 CURED PHYSICALS	ASTM or OTHER SPEC. / UNITS	GRADE W	GRADE X	GRADE Y	GRADE Z	GRADE MM	GRADE NN	GRADE SS	GRADE TT
Thermal Cond.	BTU/hr.ft²°F	---	---	---	---	---	---	---	---
Thermal Conductivity	W/m°C	---	---	---	---	---	---	---	---
Coeff. Therm. Exp.	10⁻⁵/°C	---	---	---	---	---	---	---	---
1.4 MECHANICAL PROPERTIES	ASTM or OTHER SPEC. / UNITS					——General 300,000 Lb/in² (2100 MPa)——			
Mod. of Elast, Ten. DA12	1000Lb/in²		800	1000	3000				
	MPa		5.5	6.9	21				
Shear Strength, St. Pin/Collar	Lb/in²	490/150	500/---	2200/---	2400/---				
	MPa	3.4/1	3.8/---	15/---	17/---				
Shear Strength, Nut & Bolt	Lb/in					170/160	220/280	2200/3300	610/1300
—Break/Prevail—St. Degreased	MPa					1.2/0.8	1.5/1.9	15/23	4.2/9.0
Shear Strength, Nut & Bolt	Lb/in²					1700/800	2200/700	3100/2800	2700/1400
—Break/Prevail—Zn. Phos.						12/5.5	15/4.8	21/19	19/9.6
Compressive Str.Thin Film	1000Lb/in²	125	---	---	175	---	---	---	---
	MPa	860	---	---	1200	---	---	---	---

TABLE 2.1 (Continued)

	ASTM or other spec.	UNITS	TYPE V GRADE W	TYPE V GRADE X	TYPE V GRADE Y	TYPE V GRADE Z	TYPE VI GRADE MM	TYPE VI GRADE NN	TYPE VI GRADE SS	TYPE VI GRADE TT
Commercial Number (Loctite Corp.)			567	518	510	660	203	202	204	201
1.5 ELECTRICAL PROP.										
Dielec. Str.	D149	V/mil	—General 250—				—General 250—			
1.6 HEAT/COLD RESISTANCE		UNITS								
General Rating		°F(°C)	400(204)	250(121)	400(204)	300 (149)			300(149)	400(204)
Hot Str.@ Rated Temp.			14	—	—	56	24	35	57	74
Heat Aged at Rated Temp. % Ret.			Pipe th'd	—	—	P&C	—1000 Hours break/prevail—	—1000 Hours break/prevail—	—1000 hours Break/prevail—	—1000 hours Break/prevail—
			21	—	—	131	86/100	80/126	50/115	5/33
Cold Strength—Ref. 72°F(22°C) Air			—	—	—	100	100	100	100	—
Minus 100°F(-73°C)-Acetone+CO$_2$—%			—	—	—	—	—	102	100	—
Minus 320°F(-196°C)—Liq.N$_2$—%			—	—	—	—	90	90	—	—
1.7 CHEMICAL RESISTANCE 188°F(87°C)			4 Wks	15 Days		30 Days at 188°F(87°C)			30 Days	
Air Reference——Mil-S-22473D-%			100	100		100	100	100	100	100
Motor Oil——%			100	109		120	100	100	93	88
Water——%			64	66		54	56	110	120	110
Glycol/Water (50/50%)——%			87	—		—	100	95	110	120
Transmission Fluid——%			97	98		—	40	—	90	—
Gasoline——%			77	51		110	69	100	90	94
Skydrol——%			—	—		—	—	—	100	—
Gasohol——%			—	—		—	—	—	—	—
Trichloroethane——%			100	5		120	—	—	—	—
Butyl Alcohol——%			—	—		74	64	100	—	—
Phosphate Ester——%			110	79	95	120	110	110	100	100
Toluene——%			100	80	17	—	61	100	100	88
Isopropyl Alcohol——%			—	74	8	—	—	—	94	94
1.8 MILITARY SPECIFICATION			—			—	—	—	—	—

TABLE 2.2   Proof Load of Steel Bolts (English)

SAE grade		Minimum tensile (lb/in.^{2})	Proof load[a] (lb/in.^{2})
2	1/4 to 3/4 in.	74,000	55,000
	7/8 to 1 1/2 in.	60,000	33,000
5	1/4 to 1 in.	120,000	85,000
	1 1/8 to 1 1/2 in.	105,000	74,000
8	Up to 1 1/2 in.	150,000	120,000

[a]75% of the proof load is commonly used as the working load for computing the tightening torque. Many other grades and materials are available. See Fig. 5.2 and check with the bolt supplier.

TABLE 2.3   Stress Area of Threads (English)

Bolt size (nominal in.-threads/in.)	Stress area (in.^{2})	Bolt size (nominal in.-threads/in.)	Stress area (in.^{2})
1/4–20	0.0317	9/16–12	0.1816
1/4–28	0.0362	9/16–18	0.2026
5/16–18	0.0522	5/8–11	0.2256
5/16–24	0.0579	5/8–18	0.2555
3/8–16	0.0773	3/4–10	0.3340
3/8–24	0.0876	3/4–16	0.3724
7/16–14	0.1060	7/8–9	0.4612
7/16–20	0.1185	7/8–14	0.5088
1/2–13	0.1416	1–8	0.6051
1/2–20	0.1597	1–14	0.6791

TABLE 2.4   Proof Loads of Steel Bolts
(Metric)

Class[a]	Minimum tensile (megapascal)	Proof load[a] (megapascal)
4.6	400	240
8.8	830	660
9.8	900	720
10.9	1040	940
12.9	1200	1100

[a]The metric class numbering system uses the
ultimate strength as the first digit(s) in 100s
of MPa and the first digit after the decimal
point as the minimum yield (or proof) load in
percent of tensile, e.g., Class 4.6 is a 400
MPa steel with a minimum yield of 0.6 × 400 =
240 MPa.

Torque Augmentation

   Normal loosening torque of a Unified National Coarse Threaded (UNC)
bolt will be 70% ± 10% of the torque to which it has been tightened
(UNF = 80%).

   The application of a machinery adhesive adds to the normal loosening torque. The amount by which it does this is called torque augmentation. This is shown in the shaded area in Fig. 2.1. The torque value
of augmentation is related to the breakaway torque* and may vary
between 70 and 140% of the breakaway. For the compounds shown
the break is essentially equal to augmentation.

   Most structural fasteners are torqued to at least 75% of their minimum yield strength (proof load). To prevent damage to a secured bolt
during removal, a locking material should be used that has an augmentation or breakaway that would make the break-loose torque roughly
equal to the tightening torque.

   As a design rule, to prevent shearing on removal select a material
so that:

Breakaway = 30% of tightening torque

*Breakaway torque is the torsional strength of adhesive on an un-
torqued bolt (e.g., pretorque = 0).

TABLE 2.5   Stress Area of Threads (Metric)

Nominal size (mm)	Stress area (mm^2)	Nominal size (mm)	Stress area (mm^2)
M 2.0	2.1	M 10.0	58.0
M 2.5	3.4	M 12.0	84.3
M 3.0	5.0	M 14.0	115.0
M 3.5	6.8	M 16.0	157.0
M 4.0	8.8	M 20.0	245.0
M 5.0	14.2	M 24.0	353.0
M 6.0	20.1	M 30.0	561.0
M 6.3	22.6	M 36.0	817.0
M 8.0	36.6	M 42.0	1120.0

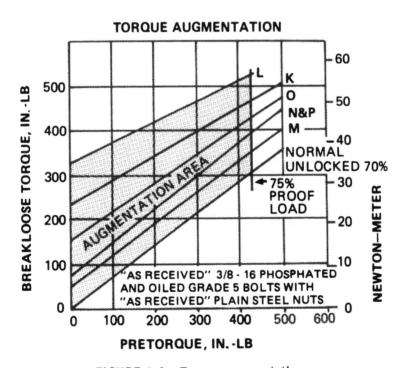

FIGURE 2.1   Torque augmentation.

TABLE 2.6 Torque Coefficient K[a]

	Oil only	Type I		Type II			Type III	
		K	L	M	N	O	R	S
Lubricated with 5% Solution of Soluble Oil (Heat Bath Corp. #72D)								
Steel	0.15	0.23	0.20	0.16	0.14	0.13	—	0.23
Phosphate	0.13	0.12	0.14	0.14	0.11	0.10	0.16	0.26
Cadmium	0.14	0.13	0.14	0.12	0.13	0.13	—	0.19
Stainless 404	0.22	0.18	0.18	0.21	0.17	0.14	—	—
Zinc	0.18	0.17	0.20	0.16	0.16	0.13	0.19	0.30
Brass	0.16	0.22	0.15	0.14	0.09	0.10	—	0.30
Silicon bronze	0.18	0.25	0.24	0.15	0.24	0.17	—	—
Al. 6262-Ta	0.17	0.21	0.25	0.20	0.29	0.18	—	—
Black oxide	0.17	0.23	0.20	0.19	0.20	0.15	0.21	0.21
Ti 6Al 4V	0.33	—	—	—	—	—	—	0.36
Degreased Fasteners								
Steel	0.20	0.22	0.26	0.18	0.20	0.18	—	0.29
Black oxide	0.40	—	—	—	—	—	—	0.19
Phosphate	0.22	0.19	0.20	0.15	0.14	0.11	—	0.28
Brass	0.26	—	—	—	—	—	—	0.28
Nylon	0.05	0.18	0.12	0.15	0.13	0.15	—	—
Zinc	0.38	0.23	0.23	0.17	0.17	0.12	0.24	0.34
Stainless 18-8	0.17	0.20	0.19	0.13	0.16	0.10	0.19	—
Cadmium	0.20	0.20	0.17	0.15	0.19	0.12	0.20	0.20
Ti 6Al 4V	0.35	—	—	—	0.25	0.22	—	0.32
Ti commercial pure	—	—	—	—	—	0.16	—	—

[a] Accuracy of the above results can vary depending on the contact area, thread form, finish, oxide contamination, and place the application. In all cases the nut was turned against a soft washer. To insure consistency of torque with the liquid materials, the threads and the thrust face were both

Type IV		Type V				Type VI			
T	U	W	X	Y	Z	MM	NN	SS	TT
0.21	—	0.13	0.23	0.26	—	—	—	—	—
0.17	0.18	0.08	0.18	0.19	0.18	0.11	0.13	0.18	0.15
0.14	—	0.12	0.20	0.20	—	—	—	—	—
0.18	—	—	—	—	—	—	—	—	—
0.18	—	0.14	0.26	0.33	—	—	—	—	—
0.20	—	0.15	0.22	0.23	—	—	—	—	—
0.27	—	—	—	—	—	—	—	—	—
0.28	—	—	—	—	—	—	—	—	—
0.21	—	0.13	0.23	0.25	—	—	—	—	—
—	—	0.16	0.28	0.30	—	—	—	—	—
0.28	0.20	0.13	0.28	0.31	—	—	—	—	—
—	—	0.11	0.20	0.28	—	—	—	—	—
0.20	—	—	—	—	—	—	—	—	—
—	—	0.14	0.29	0.33	—	—	—	—	—
0.16	—	—	—	—	—	—	—	—	—
0.23	0.20	0.14	0.29	0.33	—	—	—	—	—
0.22	—	—	—	—	—	—	—	—	—
0.20	—	0.09	0.20	0.19	—	—	—	—	—
0.23	—	0.16	0.28	0.31	—	—	—	0.22	—
—	—	—	—	—	—	—	—	—	—

covered with material. With Type VI dry materials, only the external threads were filled with material, and normal processing included a Lab oil overcoat. Nuts were oily or dry according to the chart. Products M, N, and O are especially formulated to give close control of lubricity on lightly oiled parts.

To find the correct threadlocker, enter the performance charts shown in Figs. 2.2 and 2.3 with 30% of your bolt tightening torque.

2.2   Severity of Service

Although the strength of the adhesive bond is not directly responsible for securing against self-loosening, it is an indication of the modulus and ability to resist thread sliding motion. For all Grade 5 and 8 (8.8 and 10.9) fasteners the strongest material is generally used because the use of these strong bolts usually indicates severe duty. Care must be taken to adjust the torques for thread engagement if standard nuts are not used. The breakaway torque is directly proportional to the length of thread engagement up to 3XD. Even the strongest bolt may not disassemble properly in a long tapped hole.

2.3   Selection of Viscosity to Assure Thread Filling

Select a viscosity that will apply easily, not run off, and will fill the maximum clearance (Table 2.7).

2.4   Consideration of the Cure Speed Needed

If quality control checks or functional stresses are to be applied soon after assembly, be sure that material has cured enough to avoid bond failure. A discussion of cure speed and breakaway torques vs. time can be found in Sect. 3 in this chapter.

2.5   Selection of Application Method

The method of application may dictate the particular grade selected. Application methods covered in Chap. 4 vary from full off-line automation with preapplied Grades MM, NN, SS, and TT to using a hand applicator to wick Grades R and S. For complex applications you will benefit by working closely with a supplier who has system capability.

2.6   Reusability

Preapplied materials MM, NN, SS, and TT will pass the Industrial Fasteners Institute* Specification IFI 124 for five reuses. Other machinery adhesives will not. In the author's experience this specification is too low to prove that the threads are completely filled.

---

*1505 East Ohio Building, 1717 East 9th Street, Cleveland, OH 44114.

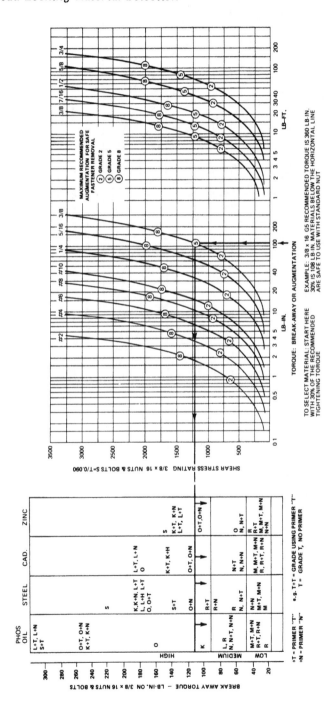

FIGURE 2.2 Thread locking performance chart (English). (See construction notes in Appendix, p. 77.)

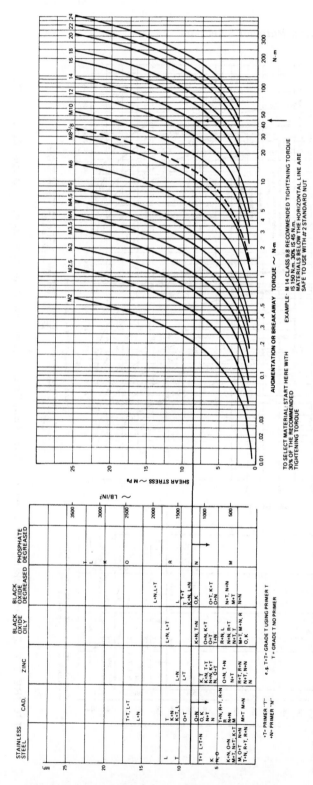

FIGURE 2.3 Thread locking performance chart (metric). (See construction notes in Appendix, p. 77.)

The safest way to reuse a fastener with any locking material or device is to apply a thread locker M, N, or O over the reused one. When reusing a machinery adhesive-locked fastener clean the threads by blowing off loose material and rinse with a squirt of activator. The same treatment will restore a nylon patch, crimped thread, or lock washer to better performance than when new.

3. VARIABLES AFFECTING CURE SPEED AND INITIAL STRENGTH

3.1 Gap or Volume Cured

In their containers, anaerobic machinery adhesives are stabilized by the presence of air, which permeates the bottles and the liquid by molecular movement. To make these adhesives cure and crosslink, the air must be excluded so that metallic ions can start the process. Since any given surface will have only a set number of ions to overbalance the dissolved oxygen it follows that the lowest volume-to-surface ratio will cure most readily. Large volumes of material can passivate an otherwise active surface. Thus, if parts are disassembled before or after cure they should be reactivated with primer T or N.

Bondlines of 0.001 to 0.003 in. (0.025 to 0.075 mm) are ideal for rapid cure and maximum strength. In general, cures will be effective up to 0.010 in. (0.25 mm) with about four times longer fixturing and 60% of the strength.

Average gap or bondline thickness does not tell the whole story because fixturing can occur where parts are touching even though the maximum clearance is larger than ideal. Gap cures have been tested in specimens with controlled gaps. Pressure retention was used as the criterion of cure (Figs. 2.4 and 2.11).

3.2 Presence of Air

The distance of the material from the closest air surface influences the cure speed. Incomplete fill (trapped air bubble) is the most common cause of slow or noncure. It is important to avoid air entrapment and to provide excess material at the juncture between the parts.

The shape of the mating surfaces has a small influence on gap curing. A 3/8 × 16 bolt and nut assembly has intimate contact on the loaded thread flank, leaving the unloaded flank in clearance. The helical gap thus formed is 5 in. (127 mm) long, so that most of the volume of material is remote from any air surface. This special configuration of a thread creates more tolerance to gap cure than does a cylindrical fit, where the air travel is strictly longitudinal. Fortunately, machinery adhesives can overcome their anaerobicity if given a little more time for cure.

TABLE 2.7 Viscosity and Thread Clearance

Grade	Viscosity (cP or mP.s)	Strength, relative	Suggested bolt range		Maximum diametral clearance		Equivalent preapplied grade[a]
			(in.)	(mm)	(in.)	(mm)	
M	1000	Low	#2–1/2	2.2–12	0.016	0.4	MM
N	1100	Medium	1/4–3/4	6–20	0.022	0.6	NN
O	1500	High	3/8–1	10–24	0.025	0.6	NN, SS, or TT
K	750	High	3/8–1	10–24	0.025	0.6[b]	SS or TT
L	7000	High	5/8–1+	16–24+	0.025	0.6[b]	SS or TT
R	12	High	#2–1/2	2.2–12	0.016	0.4[b]	SS or TT

[a]Grade SS is suggested for adhesion to plated fasteners.
[b]Primer/activator T is recommended for curing in the maximum gaps.

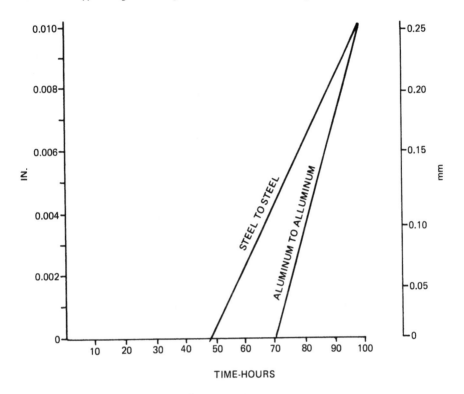

FIGURE 2.4   Gap vs. cure time, Grade Y.

3.3   Active or Inactive Surfaces

Activity of the surface will also influence curing results. The thinner
grades can have their effective gap cure doubled with the presence of
activator T or N on the surface. In general, the chemical activation of
the surface will speed the cure, but sometimes the trade-off is lower
strength. The cure speed graphs for the effect of various surfaces
should be carefully observed. Data are not available on all grades and
all surfaces. More surfaces are shown in Fig. 1.19 for Grade N. This
bar graph shows relative strengths on a wide variety of surfaces;
however, one should not assume the identical relationship for other
materials. Individual tests should be conducted. In general, active
and inactive surfaces are divided as follows (see also Figs. 2.5–2.13):

Active

Zinc phosphate	Titanium (6A1 4V and commercially pure)
Steel or iron	
Copper or brass	Magnesium alloys
Aluminum (commercial)	Nickel
	Manganese

Inactive or slow

Zinc	Dacromet steel
Cadmium	All thermoset plastics
Zinc dichromate	Stainless steel
Polyseal	Aluminum (pure)
Anodized or passivated surfaces	Glass
Magnesium (pure)	Glass epoxy
Gold, silver	Ceramics
Platinum	

Data for Fig. 2.11 were obtained on 0.375 in. (9.5 mm) thick steel flanges with gaps controlled between the mating surfaces at zero, 0.005 in. (0.13 mm), and 0.010 in. (0.25 mm). Test pressures were limited to 300 psig (2.7 MPa) in a fixture as illustrated. High-pressure test fixtures (not shown) could be sealed at 1000 psi with a 0.05 in. (1.3 mm) gap, and cure was assured by using accelerator/primer N and a 48-hour cure time. (See Figs. 2.12–2.14.)

3.4   Primer/Activator N and T

Primers N and T are surface preparatory rinses that provide mild degreasing action and accelerate normal room temperature cures of all of the machinery adhesives. In general, use the first one mentioned in Table 2.1. Primers T and N are used to assure cure on inactive surfaces or at reduced temperatures and to remove variability of cure time from active surfaces. On the thinner grades they will double the gap curing ability. Repair and maintenance operations on dirty materials of unknown composition should include a rinse with one of the activators.

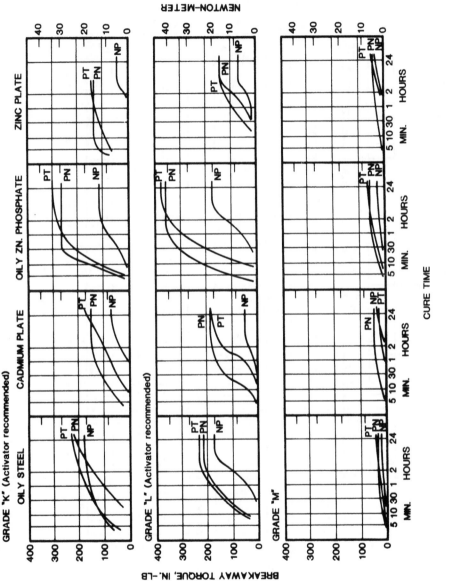

FIGURE 2.5 Cure speed on 3/8 × 16 nuts and bolts, Grades K, L, and M. PT = Primer T; PN = Primer N; NP = no primer.

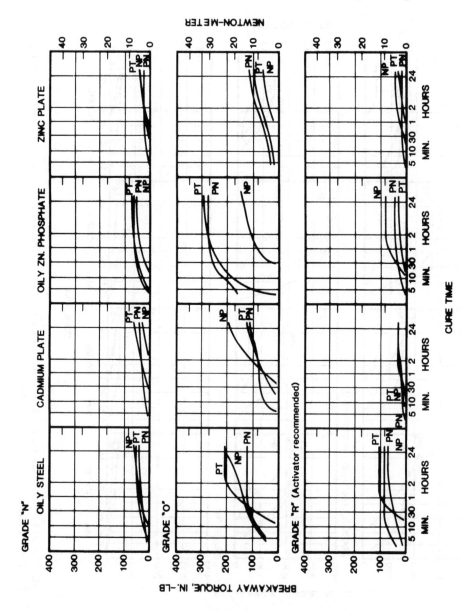

FIGURE 2.6   Cure speed on 3/8 × 16 nuts and bolts, Grades N, O, and R.   PT = Primer T;
PN = Primer N; NP = no primer.

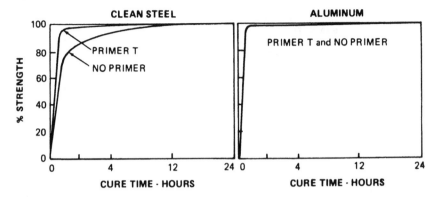

FIGURE 2.7   Cure speed on pins and collars, Grade S.

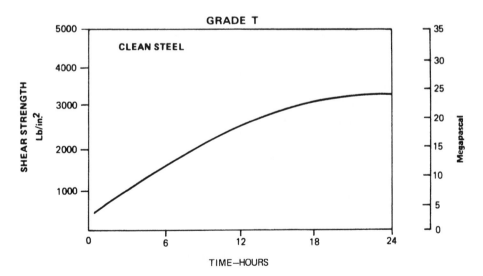

FIGURE 2.8   Cure speed on pins and collars, Grade T.

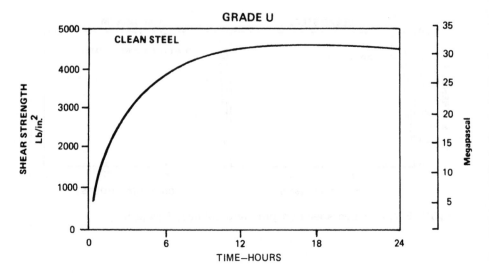

FIGURE 2.9   Cure speed on pins and collars, Grade U.

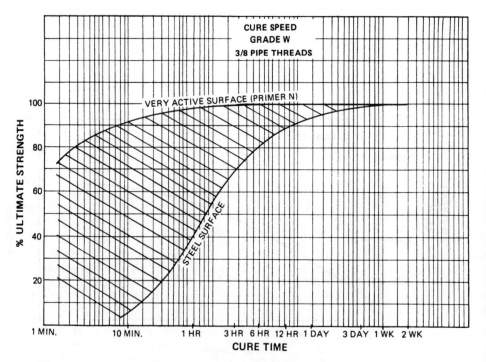

FIGURE 2.10   Cure speed on pipe threads, Grade W.

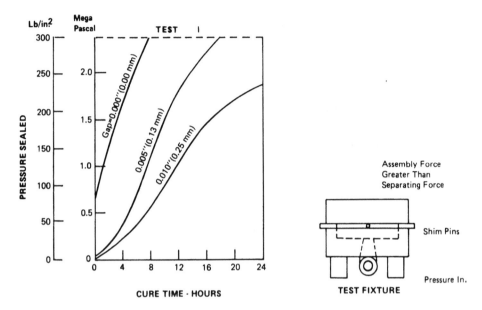

FIGURE 2.11   Cure speed on flanges, Grade X.

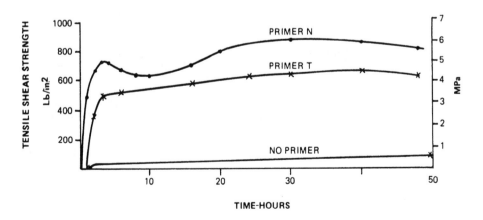

FIGURE 2.12   Cure speed on steel laps, Grade Y.

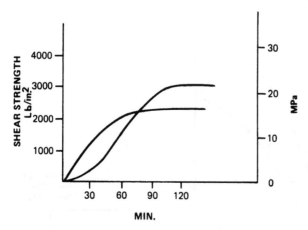

FIGURE 2.13   Cure speed on steel laps, Grade Z.

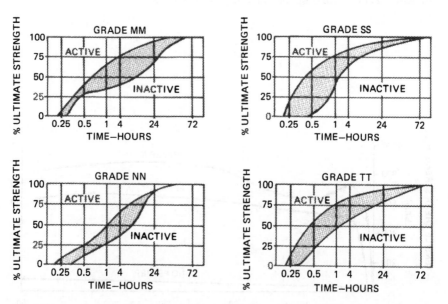

FIGURE 2.14   Cure speed on 3/8 × 16 nuts and bolts, preapplied MM, NN, SS, and TT.

Production applications usually are clean and the composition of parts known. Grades M, N, O, T, U, and Z are responsive to moderately inactive surfaces and seldom need the assistance of an activator. Activator specifications for primers N and T are as follows:

Property	Primer N	Primer T
Color	Green	Yellow
Solvent	Trichlorethane	Trichlorethane
Viscosity cP (kPa.s)	3	3
Flash point	None	138°F (59°C)
Toxicity TLV PPM	350	350
Specific gravity	1.32	1.3
Concentrate/solvent[a]	1:30 by volume	1:9 by volume
Mil. Spec.	Mil-S-22473D	–
Drying time (minutes)	3	3
On-part life (weeks)	4	1
Shelf life (unopened)	1 year	1 year

[a]Material is usually shipped ready to use but is available as concentrate.

3.5 Temperature

All chemical reactions can be speeded by elevating the temperature, but there are limits above which molecules do not combine; instead they tear themselves apart. For machinery adhesives the upper limit is about 325°F (163°C). Above this temperature some of the constituents will evaporate before they cure. As a rule of thumb, small parts assembled with machinery adhesive will cure completely given the following oven times and temperatures. Included is a 10-minute soak to get the bondline up to temperature.

Temperature °F (°C)	Oven time (minutes)
300°F (149°C)	15
250 (121)	30
200 (93)	60
150 (66)	180

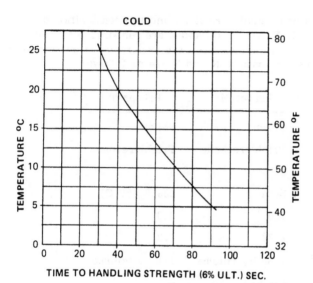

FIGURE 2.15   Effect of temperature on cure.

Induction heat, which gets the bondline up to temperature in seconds, can fixture parts in 15 seconds and produce substantial strength in 30 seconds as the parts cool. As with rapid activator curing, the forcing of cure with heat may not give ultimate strengths as high as that given by slower cures, which allow the molecules to adjust to the slight shrinkage (about 6%) that always occurs.

Lowering the temperature reduces the curing action in a faster than linear relationship (Fig. 2.15). Grades M, N, and O will cure into an unusable, mushy solid at 0°F (−18°C) in 24 hours. They must be warmed to room temperature to complete the cure, which will happen if they are heated within a few days. Thicker materials will be almost solid below 0°F and cannot be properly applied.

Preapplied Grades MM, NN, SS, and TT have built-in activators that energize on assembly. The comparison of 36 and 72°F cures for Grade SS are shown in Fig. 2.16.

More data on curring at lowered temperatures are in Chap. 3, Sec. 4.1.

3.6   Humidity

The effects of humidity are to increase the cure speed and decrease the final strength. When parts and adhesive are stabilized at 100% relative humidity, the strength suffers by as much as 40%. Tests for speed and strength should be done at 50% RH.

3.7   Finish on the Parts

Tests show that rougher finishes give higher break strengths, if the
break is perpendicular to the lay of the finish. Between finishes of
10 and 120 microinch (0.3 and 3 micrometer) the gain is 1200 lb/in.[2]
(8 MPa) and linear, regardless of the strength level of the material.
If the break is made parallel to the lay, such as torquing a lathe-
turned pin and collar instead of pushing it out, the rougher finish
gives lower strength and the loss exactly equals the cross-lay gain.
In other words, there is no effect on strength from changing the finish if the roughness is directionally oriented and the break is made
parallel to the orientation. Gains in strength are made only when the
roughness increase is perpendicular to the stress. Because functional
stresses are usually multidirectional, it is best not to increase the
rating of a material because of unidirectional matching roughness.
Multidirectional roughness can give universal increase in strength.

   The best roughening for multidirectional strength is one that is
obtained with a sandblast or similar multidirectional process. Others
are light diamond or straight knurls, hand abrading with emery cloth,
and abrasive tumbling.

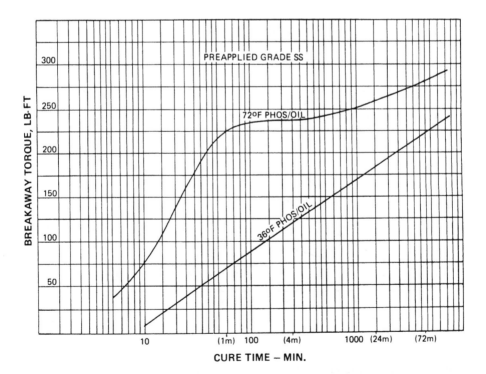

FIGURE 2.16   Cure rates of Grade SS at 36 and 72°F.

4. SURFACE COVERAGE AND QUANTITY TABLES

Figs. 2.17–2.21 and Tables 2.8–2.15 are to be used for estimating usage of material in a proposed application. Once production starts the quantities can be fine-tuned to the actual tolerances of fit. In treating nuts and bolts the estimates take into account the normal tolerances, which have a volume-fill variation of 8:1. Because of this variation in production parts, an excess of material around the last thread is the preferred average situation. Slip fitted parts normally average about half as much variation or 4:1.

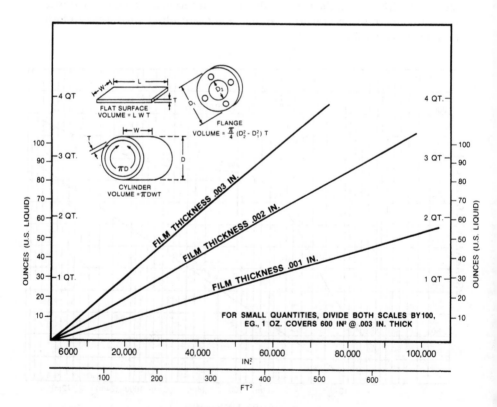

FIGURE 2.17   Surface coverage (U.S. liquid oz to in.^{2} and ft^2).

TABLE 2.8 Activating Areas Covered by Primer/Activators N and T

Package size[a]	Area covered		
	(in.^{2})	(cm^2)	(m^2)
4 oz (vol.) bottle	3,600	23,000	2.3
6 oz (wt.) spray can	5,200	34,000	3.4
32 oz (qt, vol.) can	29,000	190,000	19
1 gal can	120,000	770,000	77

[a]Conversion factors: 16 oz (avoirdupois or wt.)/lb, 16 oz (vol.)/pt, 32 oz (vol.)/qt, 128 oz (vol.)/gal, density N and T 1.4 oz (wt.)/oz (vol.).

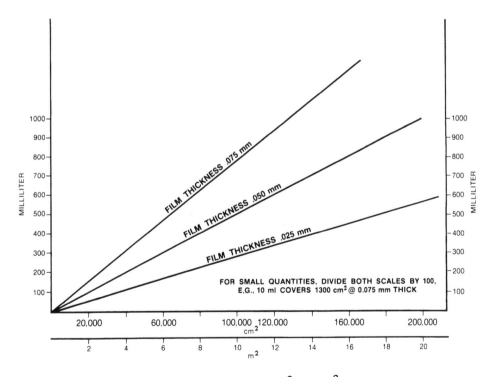

FIGURE 2.18 Surface coverage (ml to cm^2 and m^2).

TABLE 2.9   Treating Nuts and Bolts (Number of Milliliters Required
to Treat 1000 Pieces)

Bolt size		Manual application from bottle	Application with equipment
(in.)	(nominal size)		
1/4	M 6	27	17
5/6	M 9	45	30
3/8	M 10	75	50
7/16	—	105	70
1/2	M 12	135	90
9/16	M 14	180	120
5/8	M 16	225	150
3/4	M 20	340	230

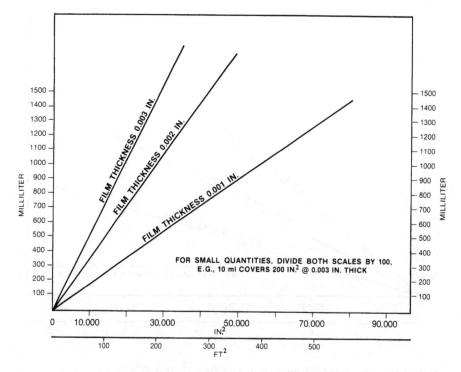

FIGURE 2.19   Surface coverage (ml to in.^{2} and ft^2).

TABLE 2.10 Treating Pipe Fittings (Number of Milliliters Required to Treat 1000 Pipe Fittings

Pipe size			
(in.)	(mm)	Tumble	Manual
1/8	6	25	40
1/4	8	45	60
3/8	10	60	90
1/2	15	90	130
3/4	20	190	250
1	25	360	440

TABLE 2.11 Measuring a "Drop" (Number of Free-Fall Drops per Milliliter from a Pointed Nozzle)

Viscosity (cP or mPa.s)	Drops/ml
1–100	100
100–1000	70
1000–5000	50
5000–10,000	30

TABLE 2.12 Sealing Welds

Low-viscosity Grade R will cover 100 linear in./ml (250 cm/ml) of weld when applied with a 1/2 in. brush (about 63 in.^{2}/ml or 390 cm^2/ml).

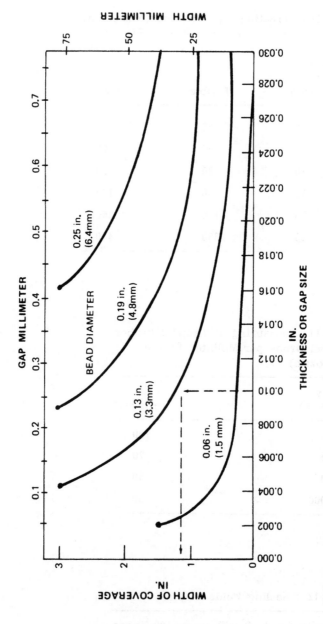

FIGURE 2.20   Bead size and area covered—pastes.

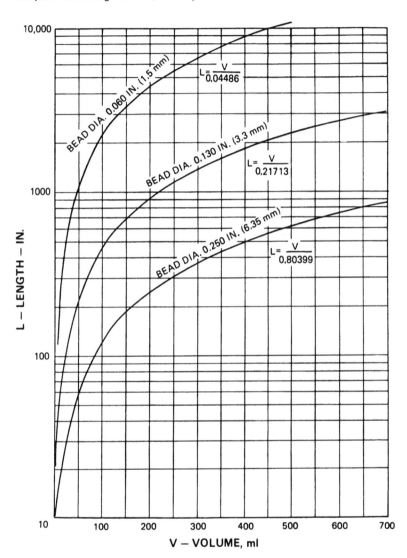

FIGURE 2.21   Bead length vs. volume (in. to ml).

TABLE 2.13   Preapplied Resin-Slurry Usage on Bolts:
Type VI Grades MM, NN, SS, and TT Applied as a
Water-Based Slurry

Bolt size		Grams per 1000 fasteners[a]
(# or in.)	(nominal size)	
#2	M 2.2	14
#4	M 3	22
#6	M 3.5	29
#8	M 4	37
#10	M 5	56
1/4	M 6	77
5/16	M 9	120
3/8	M 10	175
7/16	–	250
1/2	M 12	335
9/16	M 14	450
5/8	M 16	590
3/4	M 20	900
7/8	M 22	1300
1	M 24	1750
1 1/8	M 30	2350
1 1/4	M 33	3100
1 3/8	M 36	4100
1 1/2	M 39	5400

[a]Values are based on machine treating fasteners with
a band width 1.5 × diameter (1 lb =434 g).

TABLE 2.14 Preapplied Resin-Slurry Usage on Pipe
Fittings: Type VI Grades MM, NN, SS, and TT Applied
as a Water-Based Slurry

Pipe size		Coating width		
(in.)	(mm)	(in.)	(mm)	ml/1000 parts[a]
1/16	—	0.3	8	30
1/8	6	0.4	10	50
1/4	8	0.4	10	112
3/8	10	0.5	13	155
1/2	15	0.6	15	285
3/4	20	0.6	16	400
1	25	0.8	19	725
1 1/4	32	0.8	19	980
1 1/2	40	0.9	22	1150
2	50	1.1	27	1530
2 1/2	65	1.1	29	3200
3	80	1.3	32	4000

[a]480 ml per lb.

TABLE 2.15 Tumbling of Screws: Milliliters Required to Coat 1000 Pieces of Round-Head or Socket-Head Cap Screws with Grade W (Do Not Activate the Screws)

Length (in.)	(mm)	#2 2.2	#3 2.5	#4 —	#5 3	#6 3.5	#8 4	#10 5	#12 —	1/4 6	5/16 9	3/8 10	1/2 12
							Diameter (in. and mm)						
1/8	3	2	2	—	3	4	—	—	—	—	—	—	—
1/4	6	3	3	3	5	6	7	8	10	13	—	—	—
3/8	10	4	4	4	6	7	9	10	13	16	22	—	—
1/2	13	5	6	5	7	9	10	13	15	18	25	32	—
3/4	19	7	8	6	10	12	14	17	20	24	32	41	55
1	25	9	10	9	13	15	18	21	25	30	38	49	66
1 1/2	38	—	—	10	19	21	25	30	34	41	53	66	87
2	51	—	—	16	24	28	33	38	44	52	65	83	110
3	76	—	—	—	—	—	47	55	63	75	94	120	140

APPENDIX
Performance Chart Construction Notes

Figs. 2.2 and 2.3 were laid out to compare shear stress with torque
for bolt sizes #2 to 3/4 in. They not only compare bolt sizes but also
are compensated for the increased shear strength of materials on
thinner bond lines (small screws).

Each curve is calculated from the shear stress empirically found
for each material on 3/8 × 16 bolts and nuts. The formula used was:
Torque = stress × area × radius. Cure strengths were determined
after 24 or 72 hours at room temperature or 200°F, whichever were
highest. Stresses determined on pins and collars per Mil-R-46082 are
usable, although they will average somewhat higher because of the
smaller clearances.

$$T = S \times (\pi D_p \times 2Le) \times D_p /2 \text{ or } \pi S (D_p)^2 Le$$

where S = shear stress; D_p = pitch diameter; Le = engaged length,
which for a standard nut is nominal diameter × 0.8.

The shear stress for each diameter of screw was determined experimentally. The performance curves were skewed in accordance with
the shear-stress vs. clearance graph in Fig. 1.18.

Chapter 3
Environmental Effects

1. SOLVENT RESISTANCE

1.1 General

Predicting the suitability or life of any material for a particular application or environment is difficult without extensive field tests that duplicate the proposed environment. This is especially true for adhesive systems because the adhesive is only half of the system. The substrates and their preparation are as important as the adhesive. Fortunately, the anaerobic adhesives are highly crosslinked thermoset plastics, which are extremely difficult to unlink. They are resistant to most hydrocarbons (oils, gasoline), chlorinated solvents, water, mild acids, and alkalis.

Machinery adhesives, like any durable plastic, can be made to fail by any one or a combination of mechanisms. Examples of such mechanisms are: molecular breakdown by strong chemical reaction, solvation, absorption, stress cracking, mechanical stressing, delamination of adhesion, and desorption, all of which are made more rapid by elevated temperatures. It is rarely necessary to test these phenomena individually because machinery adhesives usually are confined within metal parts that protect the adhesive from exposure except for a very thin bondline. The exposure of only a thin bondline suggests the most important requirement for long-term chemical resistance. That is, the joint must be completely filled so that the environmental chemical cannot penetrate and reside in voids in the joint.

Such voids increase the exposed area as well as extend the time of contact in many situations. After a rainstorm stops, the wet void remains as a corrosive pocket long after the external parts are dry. Often adhesion and strength can be recovered if the joint is only intermittently exposed. The most common failure is separation at the interface, which means that shear strength is lost but reasonable sealing may last considerably longer.

Field experience has shown that severe weathering of construction equipment left outdoors on site or in open or underground mines has not affected the adhesive joints. There are examples of flawless service recorded for over 10 years. Fasteners in submarines have been sealed and secured for over 15 years of continuous duty. Outdoor railings assembled with early versions of machinery adhesives are still giving excellent service after 20 years in New York City.

Since field experience in all conditions is never available it would be nice to have laboratory tests that one could confidently say would predict future performance. Such tests have been developed by the theorists and they do prove to be conservatively predictive *for the conditions that they impose.*

1.2 Hot Solvent Tests

Resistance to commercial solvents and oils can be predicted by temperature-accelerated ageing. The rule of thumb used by chemists is that every increase of 10°C in temperature will double the effect on organic molecules. Whether or not this holds true for all types of material and degradation (e.g., dissociation and adhesion release) is unknown, but experience indicates that it gives a conservatively based life prediction for machinery adhesives.

Many tests have been done at 87°C (188°F) which condenses 8 years of degradation into 1 month. According to the rule of thumb just cited, there are 6 1/2 doublings of time between 22°C (72°F) and 87°C (188°F), or $2 \times 2 \times 2 \times 2 \times 2 \times 2 \times 1.5 = 2^6 \times 1.5 = 64 \times 1.5 = 96$ factor. Therefore one month at temperature is equivalent to

1 month/12 months/1 year × 96 = 8 years

Materials that experience temperatures below 22°C, of course, have their aging slowed and their life extended.

1.3 Compatible Chemicals

Table 3.1 is a guide for the selection of anaerobic materials for sealing and locking in the presence of liquids and gases. It also indicates the degree of suitability. Ratings were selected on the basis of limited field tests and the knowledge of chemical activity. Testing is always recommended because the substrate is as important as the adhesive. Hot water affects copper far more than it does stainless steel regardless of the bonding material.

TABLE 3.1   Compatibility of Loctite with Other Materials

LIQUIDS

Abrasive Coolant	1	Hyposulphite	1
Acetaldehyde	1	Iodide	1
Acetate Solvents	1	Molybdate	1
Acetamide	1	Nitrate	1
Acetic Acid 10%	2	Oxalate	1
Acetic Acid 80%	3	Persulphate	1
Acetic Acid — Glacial	3	Phosphate	1
Acetic Anhydride	3	Picrate	1
Acetone	1	Sulphate	1
Acetyl Chloride	2	Sulphide	1
Acetyl Salicylic Acid	1	Thiocyanate	1
Acetylene (Liquid Phase)	1	Amyl Acetate	1
Acid Clay	1	Amyl Amine	1
Acrylic Acid	1	Amyl Chloride	1
Acrylonitrile	1	Aniline	1
Activated Alumina	1	Aniline Dyes	1
Activated Carbon	1	Animal Fat	1
Activated Silica	1	Anodizing Bath	1
Albumin	1	Antibiotic Broth	1
Alcohol-Allyl	1	Antimony Chloride Solution	1
Alcohol-Amyl	1	Antimony Acid Salts	1
Alcohol-Benzyl	1	Antimony Oxide	1
Alcohol-Butyl	1	Antioxidant Gasoline	1
Alcohol-Ethyl	1	Apple juice, cider	1
Alcohol Furfuryl	1	Aqua Regia	NO
Alcohol Hexyl	1	Argon	1
Alcohol Isopropyl	1	Armeen	1
Alcohol-Methyl	1	Arochlor	1
Alcohol-Propyl	1	Aromatic Gasoline	1
Alum-Ammonium	1	Aromatic Solvents	1
Alum-Chrome	1	Arsenic Acid	1
Alum Potassium	1	Asbestos Slurry	1
Alum-Sodium	1	Ash Slurry	1
Alumina	1	Asphalt Emulsions	1
Alkazene (AR Bromo Benzine)	3	Asphalt Molten	1
Aluminium Acetate	1	Aureomycin	1
Bicarbonate	1	Bacitracin	1
Bifluoride	1	Bacterial Media	1
Chloride	1	Bagasse Fibres	1
Sulphate	1	Barium Acetate	1
Ammonia Anhydrous	3	Carbonate	1
Ammonia Solutions	3	Chloride	1
Ammonium Bisulphate	1	Hydroxide	2
Borate	1	Sulphate	1
Bromide	1	Battery Acid	2
Carbonate	1	Battery Diffuser Juice	1
Chloride	1	Bauxite (See Alumina)	1
Chromate	1	Beef Extract	1
Cupro Formate	1	Beer	1
Fluoride	1	Beet juice or pulp (or sugar liquors)	1
Fluosilicate	1	Bentonite	1
Formate	1	Benzaldehyde	1
Hydroxide	1		
Hydroxide, Nitrate Sol	1		

TABLE 3.1 (Continued)

Benzene	1	Nitrate	1
Benzene Hexachloride	1	Phosphate	1
Benzene in Hydrochloric Acid	1	Silicate	1
Benzoic Acid	1	Sulphamate	1
Benzo Triazole	1	Sulphate	1
Beryllium Hydroxide	1	Sulphite	1
Sulphate	1	Camphor	1
Bicarbonate Liquor	1	Cane Sugar Liquors	1
Bilge Lines	1	Cane Sugars Refined	1
Bleach Liquor	1	Carbitol	1
Bleached Pulps	1	Carbolic Acid (Phenol)	2
Blood Animal	1	Carbon Bisulphide	1
Extender	1	Carbon Tetrachloride	1
Human	1	Carbonated Beverage	1
Borax Liquors	1	Carbonic Acid	2
Bordeaux Mixture	1	Carbowax	1
Boric Acid	1	Carboxy Methyl Cellulose	1
Brake Fluids	1	Carnauba Wax	1
Brandy	1	Casein	1
Brine Alkaline	1	Casein Water Paint	1
Chlorinated	1	Cashew Oil	1
Cold	1	Caster Oil	1
Pickling	1	Catsup	1
Bromine	NO	Celite	1
Bromine Solution	3	Cellosolve	1
Butadiene	1	Cellulose Acetate	1
Buttermilk	1	Pulp	1
Butyl Acetate	1	Xanthate	1
Butyl Alcohol	1	Cement Dry/Air Blown	1
Amine	1	Grout	1
Cellosolve	1	Slurry	1
Chloride	1	Ceramic Enamel	1
Ether-Dry	1	Cereal cooked	1
Lactate	1	Ceric Oxide	1
Butyral Resin	1	Chalk	1
Butyraldehyde	1	Chestnut Tanning	1
Butyric Acid	2	China Clay	1
Cadmium Chloride	1	Chloral Alcoholate	1
Plating Bath	1	Chloramine	1
Sulphate	1	Chlorinated Hydrocarbons	1
Calcium Acetate	1	Chlorinated Paperstock	1
Bisulphate	1	Solvents	1
Carbonate	1	Sulphuric Acid	NO
Chlorate	1	Water	1
Chloride	1	Wax	1
Chloride Brine	1	Chlorine Dioxide	3
Citrate	1	Chlorine Liquid	NO
Ferro Cyanide	1	Chlorine Dry	NO
Formate	1	Chloroacetic Acid	1
Hydroxide	1	Chlorobenzene Dry	1
Hypochlorite	1	Chloroform Dry	1
Lactate	1	Chloroformate Methyl	1

TABLE 3.1   (Continued)

Chlorophyll	1	DDT Intermediates	1
Chlorosulphonic Acid	NO	Dionized Water	1
Chocolate Milk	1	Dionized Water Low Conductivity	1
Chocolate Syrup	1	Dental Cream	1
Chrome Acid Cleaning	2	Detergents	1
Liquor	2	Developer, Photographic	1
Plating Bath	2	Dextrin	1
Chromic Acid 10%	1	Dextran	1
Chromic Acid 50% (Cold)	3	Diacetone Alcohol	1
Chromic Acid 50% (Hot)	NO	Diammonium Phosphate	1
Chromium Acetate	1	Diamylamine	1
Chloride	1	Diatomaceaus Earth Slurry	1
Sulphate	1	Diazo Acetate	1
Citric Acid Dilute (Cold)	3	Dibutyl Phthalate	1
Citric Acid Dilute (Hot)	3	Dicyandiamide	1
Citrus Concentrate	1	Dielectric Fluid	1
Juices	1	Diester Lubricants	1
Clay	1	Diethyl Ether Dry	1
Coal Slurry	1	Dichlorophenol	1
Coal Tar	1	Dichloro Ethyl Ether	1
Coating Colours	1	Diethyl Sulphate	1
Cobalt Chloride	1	Diethylamine	1
Coca Cola Syrup	1	Diethylene Glycol	1
Cocktail Liquors	1	Diglycolic Acid	1
Coconut Oil	1	Dimethyl Formamide	1
Coffee Concentrate	1	Dimethyl Sulphoxide	1
Coke Breeze	1	Dioxane-Dry	1
Condensate	1	Dioxidene	1
Cooking Oil	1	Dipentene-Pinene	1
Copper Ammonium Formate	1	Diphenyl	1
Chloride	1	Distilled Water	1
Chloride-Gasoline	1	Distillery Mash	1
Cyanide	1	Distillery Slops	1
Liquor	1	Dowtherm	1
Napthenate	1	Drying Oil	1
Plating Acid Process	1	Dust-Flue (Dry)	1
Plating Alk Process	1	Dye Liquors	1
Sulphate	1	Edible Oils	1
Copperas	1	Emery-Slurry	1
Core Oil	1	Emulsified Oils	1
Corn Kernels	1	Enamel Frit Slip	1
Corn Oil	1	Enzyme Solution	1
Corn Steep Liquor	1	Epichloryhydrin	1
Corn Syrup	1	Ergosterol Solution	1
Corundum	1	Essential Oils	1
Cottonseed Oil	1	Esters General	1
Creosote	1	Ethyl Acetate	1
Creosote-Cresylic Acid	1	Alcohol	1
Cyanide Solution	2	Amine	1
Cyanuric Chloride	1	Bromide	1
Cyclohexane	1	Cellosolve	1
Cylinder Oils	1	Cellosolve Slurry	1

TABLE 3.1 (Continued)

Formate	1	Aviation	1
Silicate	1	Copper Chloride	1
Ethylene Diamine	1	Ethyl	1
Dibromide	1	Motor	1
Dichloride	1	Sour	1
Glycol	1	White	1
Diamine Tetramine	1	Gelatin-Edible	1
Face Cream	1	Emulsion	1
Fatty Acids	1	Gluconic Acid	1
Fatty Acids Amine	1	Glucose	1
Fatty Alcohol	1	Glue-Animal Gelatin	1
Ferri-Floc	1	Plywood	1
Ferric Chloride	1	Gluten	1
Nitrate	1	Glutamic Acid	1
Sulphate	1	Gluten	1
Ferro Silicon Slurry	1	Glycerine C.P.-USP	1
Ferrocene-Oil Sol	1	Lye-Brine	1
Ferrous Chloride	1	Glycerol	1
Oxalate	1	Glycine	1
Sulphate 10%	1	Glycine Hydrochloride	1
Sulphate (Sat)	1	Glycol Amine	1
Fertilizer Solution	1	Glyoxal	1
Fish Oil	1	Gold Chloride	1
Fission Wastes	1	Cyanide	1
Flavouring Syrups	1	Grain Mash	1
Flotation Concentrates	1	Granodine	1
Fluoride Salts	1	Grape Juice	1
Fluorene Gaseous or Liquid	1	Grapefruit Juice	1
Fluorolube	1	Grease-Edible	1
Fluosilic Acid	1	Lubricating	1
Flux Soldering	1	Green Liquor	1
Fly Ash Dry	1	Green Soap	1
Foam Latex Mix	1	Grinding Lubricant	1
Foamite	1	Grit Steel	1
Formaldehyde (Cold)	1	Gritty Water	1
Formaldehyde (Hot)	3	Groundwood Stock	1
Formic Acid (Dilute Cold)	1	GRS Latex	1
Formic Acid (Dilute Hot)	3	Gum Paste	1
Formic Acid (Cold)	1	Turpentine	1
Formic Acid (Hot)	3	Gypsum	1
Freon (see gases)		Hair Tonics	1
Fruit Juices, Berry, etc	1	Halane Sol.	1
Fuel Oil	1	Halogen Tin Plating	1
Fuming Nitric Acid (Red)	NO	Halowax	1
Sulphuric Acid	NO	Harvel-Transil Oil	1
Oleum	NO	Heptane	1
Furfural	1	Hexachlorobenzene	1
Gallic Acid 5%	2	Hexadiene	1
Gallium Sulphate	1	Hexameta Phosphate	1
Gamma Globulin	1	HTP (Conc. Rocket Fuel) 100%	1
Gasoline-Acid Wash	1	Hexamethylene Tetramine	1
Alk. Wash	1	Hexane	1

TABLE 3.1   (Continued)

Houghto Clean	1	Lapping Compound	1
HTH	1	Lard Oil	1
Hydrazine	1	Latex — Natural	1
Hydrazine Hydrate	1	Synthetic	1
Hydrobromic Acid	2	Synthetic Raw	1
Hydrochloric Acid	2	Launder Wash Water	1
Hydrocyanic Acid	1	Laundry Bleach	1
Hydrofluoric Acid	NO	Blue	1
Hydrogen fluoride	2	Soda	1
Hydrogen Peroxide (dilute)	1	Lead Arsenate	1
Hydrogen Peroxide (conc.)	3	Oxide	1
Hydroponic Sol.	1	Sulphate	1
Hydroquinone	1	Tetraethyl	1
Hydroxy Acetic Acid	1	Lecithin	1
Hypo	1	Lemon Juice	1
Hydrochlorous Acid	1	Lignin Extract	1
Ice Cream Mix	1	Lime Bleach	1
Ink	1	Saccharate	1
Ink in solvent-Printing, etc	1	Slaked	1
Insecticide	1	Sulphur Mix	1
Insuline Slurry	1	Linseed Oil	1
Iodine in Alcohol	1	Lithium Chloride	1
Iodine — Potassium Iodide	1	Liver Extract	1
Iodine Solutions	1	Low Wine — Raw	1
Ion Exclusion Glycol	1	Lox (Liquid O_2)	NO
Irish Moss Slurry	1	Ludox	1
Iron Ore Taconite	1	Lye	NO
Oxide	1	Magnesia Nitrate	1
Phytate	1	Slurry	1
Salts	1	Magnesium Bisulphite	1
Isobutyl Alcohol	1	Carbonate	1
Isobutyraldehyde	1	Chloride	1
Iso octane	1	Hydroxide	1
Isopropyl Alcohol	1	Sulphate	1
Isocyanate Resin	1	Magnesite Slurry	1
Isopropyl Acetate	1	Magnesite	1
Ether	1	Maleic Acid	1
Itaconic Acid	1	Anhydride	1
Jams — Jelly	1	Malt Slurry	1
Jet Fuels	1	Syrup	1
Jewellers' Rouge	1	Maltose	1
Jig Table Slurry	1	Manganese Chloride	1
Juice — fruit and vegetable	1	Sulphate	1
Kaolin — China Clay	1	Mannitol Sol.	1
Kelp Slurry	1	Mayonnaise	1
Kerosene	1	Melamine Resin	1
Kerosene, Chlorinated	1	Menthol	1
Ketone	1	Mercaptans	1
Kraut Juice	1	Mercuric Chloride	1
Lacquer Thinner	1	Nitrate	1
Lactic Acid	1	Mercury	1
Lactose	1	Methane	1

TABLE 3.1 (Continued)

Methyl Alcohol	1	Nitrofurane	1
Methyl Acetate	1	Nitroglycerine – Use at Customer's use	
Bromide	1	Nitroguanidine	1
Carbitol	1	Nitroparaffins – Dry	1
Cellosolve	1	Nitrosyl Chloride	1
Chloride	1	Norite Carbon	1
Ethyl Ketone	1	Oakite Compound	1
Isobutyl Ketone	1	Oil, Animal	1
Lactate	1	Castor	1
Orange	1	Coconut	1
Methylamine	1	Cod – Raw	1
Methylene Chloride	1	Corn	1
Milk	1	Cottonseed	1
Milk of Magnesia	1	Creosote	1
Mine Water	1	Emulsified	1
Mineral Oil White	1	Fish	1
Mineral Spirits	1	Fuel	1
Mixed Acid, Nitric/Sulphuric	NO	Linseed	1
Molasses Crude	1	Lubricating	1
Edible	1	Mineral	1
Mold Broths – Antibiotic	1	Olive – Edible	1
Monochlor Acetic Acid	1	Peanut – Crude	1
Morpholine	1	Soluble	1
Mud	1	Soya Crude	1
Muriatic Acid	2	Tall	1
Mustard Edible	1	Tung	1
Nalco.Sol.	1	Vegetable	1
Naptha	1	Oleic Acid, hot	1
Napthalene	1	Engine Oil Derd 2472	1
Naval Stores Solvent	1	Hydraulic Oil Dtd 585	1
Nematocide	1	Fuel Oil Derd 2485	1
Neoprene Emulsion	1	Oleic Acid, cold	1
Latex	1	Orange Juice	1
Nickel Acetate	1	Ore Fines – Flotation	1
Ammonium Sulphate	1	Ore Pulp	1
Chloride	1	Organic Dyes	1
Cyanide	1	Oxalic Acid – cold	1
Fluoborate	1	Ozone wet	NO
Ore Fines	1	Paint – Linseed base	1
Plating Bright	1	Water base	1
Sulphate	1	Remover – solvent type	1
Nicotinic Acid	2	Vehicles	1
Nitrana Sol.	1	Palm Oil	1
Nitration Acid	NO	Palmitic Acid	1
Nitric Acid	NO	Paper Board Mill Waste	1
Nitric Acid 10%	2	Paper Coating Slurry	1
Nitric Acid 20%	3	Paper Pulp	1
Nitric Acid Anhydrous	NO	Pulp with amum.	1
Nitric Acid Fuming	NO	Pulp with Dye	1
Nitro Aryl Sulphonic Acid	1	Pulp bleached	1
Nitrobenzene – Dry	1	Pulp bleached – washed	1
Nitrocellulose	1	Pulp chlorinated	1

TABLE 3.1   (Continued)

Groundwood	1	Polyvinyl Acetate Slurry	1
Rag	1	Chloride	1
Paper Stocks — Fine	1	Pomace	1
Paradichlorbenzene	1	Porcelain Frit	1
Paraffin Molten	1	Potable Water	1
Paraffin Oil	1	Potash	2
Paraformaldehyde	1	Potassium Acetate	1
Pectin Solution — Acid	1	Aluminium Sulphate	1
Penicillin Broth	1	Bromide	1
Pentachlorethane	1	Carbonate	1
Pentaerythritol Sol.	1	Chlorate	1
Perchlorethylene (dry)	1	Chloride--Sol.	1
Perchloric Acid	2	Chromate	1
Perchloromethyl Mercaptan	1	Cyanide--Sol.	1
Perfume	1	Dichromate	1
Permanganic Acid	NO	Ferricyanide	1
Peroxide Bleach	1	Hydroxide	NO
Persulphuric Acid	2	Iodide	1
Petrol	1	Nitrate	1
Petroleum Ether	1	Perchlorate	1
Petroleum Jelly	1	Permanganate	1
Phenol	1	Persulphate	1
Formaldehyde Resins	1	Phosphate	1
Sulphonic Acid	1	Silicate	1
Phenolic Glue	1	Sulphate	1
Phenyl Betanaphthylamine Any Alc.	1	Xanthate	1
Phloroglucinol	1	Press Board Waste	1
Phosphate ester	1	Propionic Acid	1
Phosphatic Sand	1	Propyl Alcohol	1
Phosphoric Acid 85% hot	NO	Propyl Bromide	1
Phosphoric Acid 85% cold	3	Propylene Glycol	1
Phosphoric Acid 50% hot	3	Proteins — Water Sol or Slurry	1
Phosphoric Acid 50% cold	3	Pyranol	1
*Phosphoric Acid 10% cold	1	Pyridine	1
Phosphorus Molten	1	Pyrogallic Acid	1
Phosphotungstic Acid	1	Pyrogen Free Water	1
Photographic Sol	1	Pyrole	1
Phthalic Acid	1	Pyromellitic Acid	1
Pickle Brine	1	Quebracho Tannin	1
Pickle Sol (for meat curing)	1	Quinone	1
Pickling Acid — Sulphuric	1	Quinine	1
*Phosphoric Acid 10% hot	3	Rag Stock Bleached	1
Picric acid solutions	1	Rare Earth Salts	1
Pine Oil finish	1	Rayon Acid Water	1
Pineapple Juice — edible	1	Spin Batch	1
Plasma — Blood	1	Spin Bath Spent	1
Plasma Diluent	1	Viscose	1
Polio Vaccine	1	Relish — Pickle	1
Polyacrylonitrile Slurry	1	Resorcinol	1
Polypentek	1	Riboflavin	1
Polyphosphoric Acid	2	River Water	1
Polysulphide Liquor	1	Road Oil	1

TABLE 3.1   (Continued)

Roccal	1	Chloride	1
Root Beer Extract	1	Chlorite	1
Rosin — Wood	1	Cyanide	1
In Alcohol	1	Ferrycyanide	1
Size	1	Fluoride	1
Rubber Latex	1	Formate	1
Rum	1	Glutamate	1
Safrol	1	Hydrogen Sulphate	1
Salad Dressing	1	Hydrosulphide	1
Salicylic Acid	1	Hydrosulphite	1
Saline Sol — Physiological	1	Hypochlorites	1
Salt Brine		*Hydroxide 20% (Cold)	3
Alkaline		20% (Hot)	3
Electrolytic		50% (Cold)	3
Pickle		50% (Hot)	NO
Refrig.	1	70% (Cold)	3
Salt — Sugar Pickle	1	70% (Hot)	NO
Sand — Air Blown or Slurry	1	Hypochlorite	1
Phosphatic	1	Lignosulphonate	1
Sea Coal	1	Metasilicate	1
Sea Water	1	Molten	1
Selenium Chloride	1	Nitrate	1
Sequestrene	1	Nitrite — Nitrate	1
Sewage	1	Perborate	1
Shave Lotion	1	Peroxide	NO
Shellac	1	Persulphate	1
Shower Water	1	Phosphate — Mono	1
Silica Gel	1	Phosphate — Tri	1
Ground	1	Potassium — Chloride	1
Silicon Tetrachloride	1	Salicylate	1
Silicone Fluids	1	Sesquicarbonate	1
Silver Cyanide	1	Silicate	1
Iodine-Aqu.	1	SilicoFluoride	1
Nitrate	1	Stannate	1
Size Emulsion	1	Sulphate	1
Skelly Solve, E, L	1	*Sodium Hydroxide	NO
Slate to 400 Mesh	1	Sulphide	1
Sloe Gin Concentrate	1	Sulphite	1
Soap Lye	NO	Sulphydrate	1
Stone — Air Blown	1	Thiocyanate	1
Soda Pulp	1	Thiosulphate	1
Sodium Acetate	1	Tungstate	1
Acid Fluoride	1	Xanthate	1
Alminate	1	Soft Drink Syrups	1
Arsenate	1	Solox — Denat. Ethyl	1
Benzene Sulphonate	1	Soluble Oil	1
Bicarbonate	1	Solvent Naphthas	1
Bichromate	1	Sorbic Acid	1
Bisulphite	1	Sorbitol	1
Bromide	1	Soup Stock	1
Carbonate	1	Sour Gasoline	1
Chlorate	1	Soybean Oil	1

TABLE 3.1   (Continued)

Soybean Sludge —Acid	1	Tetraethyl Lead	1
Spensol Solution	1	Tetrahydeophurane	1
Spent Cooking Liquor	1	Tetranitromethane	1
Stannic Chloride	1	Textile Dyeing	1
Starch	1	Finishing Oil	1
Starch Base	1	Printing Oil	1
Steam — Low Pressure	1	Thiocyanic Acid	1
Stearic Acid	1	Thioglycollic Acid	1
Steep Water	1	Thionyl Chloride	1
Sterilization — Steam	1	Thiophosphoryl Chloride	1
Stoddard Solvent	1	Thiourea	1
Streptomycin Broth	1	Thorium Nitrate	1
Styrene	1	Thymol	1
Styrene Butadiene Latex	1	Tin Tetrachloride	1
Sugar — Carbon Slurry	1	Tinning Sol, DuPont	1
Corn, Glucose, etc	1	Titania Paper Coating	1
Fondant	1	Titanium Oxide Slurry	1
Ion Exchange	1	Oxy Sulphate	1
Solution	1	Sulphate	1
Sulphamic Acid	1	Tetrachloride	1
Sulphan — Sulphuric Anhydride	1	T.N.T. Slurry	1
Sulphathiazole	1	Tobacco Wash Sol.	1
Sulphite Liquor	1	Toluol	1
Sulphite Stock	1	Toluene	1
Sulphonated Oils	1	Toluene Sulphonic Acid	3
Sulphones	1	Tomato Catsup (Ketchup)	1
Sulphonic Acids	2	Tomato Juice	1
Sulphur Slurry	1	Transil Oil	1
Solution in Carbon Disulphide	1	Trichloracetic Acid	1
Sulphuric Acid 0—7%	3	Trichlorethane	1
7—40%	3	Trichlorethylene	1
40—75%	3	Trichlorethylene — dry	1
75—95%	NO	Tricresyl Phosphate	1
95—100%	NO	Triethanolamine	1
Sulphurous Acid	3	Triethylene Glycol	1
Sulphuryl Chloride	1	Trioxane	1
Surphactants	1	Tung Oil	1
Synthetic Latex	1	Tungstic Acid	1
Syrup — Caramel Candy	1	Turpentine	1
Edible Candy	1	Ucon Lube	1
Taconite — Fines	1	Udylite Bath — Nickle	1
Talc — Slurry	1	Undecylenic Acid	1
Tall Oil	1	Unichrome Sol.Alk.	1
Tallow	1	Uranium Salts	1
Tankage — Slurry	1	Uranyl Nitrate	1
Tannery Wastes	1	Uranyl Sulphate	1
Tannic acid (cold)	3	Urea Ammonia Liquor	1
Taunin	1	Vaccine Serum	1
Tar & Tar Oil	1	Vacuum to 100 Microns	1
Tartaric Acid	1	Vacuum Below 100 Microns	1
Tergitol	1	Vacuum Oil	1
Terpineol	1	Vanadium Pentoxide Slurry	1

TABLE 3.1 (Continued)

Vanilla Extract	1	Wax	1
Varnish	1	Wax Chlorinated	1
Varsol — Naphtha Solv.	1	Emulsions	1
Vaseline	1	Weed Killer Dibromide	1
Vegetable Juice	1	Weisburg Sulphate Plating	1
Oils Edible	1	Wheat Gluten	1
Oils Non-Edible	1	Whey	1
Versene	1	Wine — Finished	1
Vinegar 40 grains and over	1	Whiskey	1
Vinyl Acetate Dry or Chloride Monomer	1	Whiskey Slop	1
Chloride — Latex Emulsion	1	Witch Hazel Sol.	1
Resin Slurry	1	White Liquor	1
Vitamins in Oil	1	Wood Ground Pulp	1
Waste treatment	1	Wort Lines	1
Water — Acid — Below PH7	1	X-Ray Developing Bath	1
PH7 to 8.0	1	Xylene	1
Alkaline — Over PH 8.0	1	Yeast — Slurry	1
Boiling	1	Zelan	1
Carbonated	1	Zeolite Water	1
Chlorinated over 100 PPM	1	Zinc Acetate	1
Deionized	1	Bromide	1
Desalting	1	Chloride	1
Distilling	1	Cyanide — Alk.	1
Filtered	1	Fines Slurry	1
Fresh — Drinking	1	Flux Paste	1
Heavy	1	Galvanizing	1
Mine Water	1	Hydrosulphite	1
River	1	Oxide in Water	1
Sandy	1	Oxide in Oil	1
Soft	1	Sulphate	1
Sterile	1	Zincolate	1
Water — "White" low ph	1	Zirconyl Nitrate	1
"White" high ph	1	Sulphate	

GASES

Acetylene	1	Chlorine Wet	NO
Acid & Alkali Vapours	1	Coke — Oven Gas — Cold	1
Air	1	Coke — Oven Gas — Hot	3
Amine	1	Cyanogen Chloride	1
Ammonia	1	Cyanogen Gas	1
Butane	1	Ethane	1
Butadiene Gas/Liquid	1	Ether — see Diethyl Ether	1
Butylene Gas/Liquid	1	Ethylene	1
By-Product Gas — (Dry)	1	Ethylene Oxide	1
Carbon Dioxide	1	Flue Gas	1
Carbon Disulphide	1	Freon (11-12-21-22)	3
Carbon Monoxide	1	Furnace Gas hot	3
Chloride Dry	1	Furnace Gas cold	1
Chlorine	3	Nitrogen Tetraoxide Vapour	1
Chlorine Dry	NO	Bromo Chlorodifluoromethane	3

TABLE 3.1   (Continued)

Nitrous Oxide	3	Oil-Solvent vapor	1
Gas drip oil	1	Oxygen 150 psi	NO
flue	1	Oxygen	NO
manufacturing	1	Ozone	NO
natural	1	Producer gas 50 psi	1
Helium	1	Propane	1
Hydrogen gas — cold	1	Propylene	1
Hydrogen chloride	1	Steam	NO
Hydrogen cyanide	1	Sulphur Dioxide	1
Hydrogen Sulphide wet and dry	1	Sulphur Dioxide dry	1
Isobutane	1	Sulphur Trioxide Gas	NO
Methane	1	Sulphur Trioxide Dry	1
Methyl chloride	1	Sulphuric Acid vapor	1
Natural gas dry	1	Titanium	1
Nitrogen gas	1		

Code: 1, Chemicals in this group are compatible with all grades of
anaerobics. 2, For 10% concentration or less, the liquids in this group
will be suitable with all grades. Over 10% concentration the better
grades are K, O, P, NN, and TT. 3, These chemicals will always require initial testing. Start with Grade K or P. NO, Use of organic
sealants is not recommended.

2. HOT STRENGTH

Strength at temperature is almost always determined in air in order to
find the effect of temperature alone on the molecular strength of the
material. A soak of 1 hour is given to be sure that the bondline is
fully up to temperature and the bond is broken. When parts are expected to perform only occasionally at temperature, the hot strength
is the one to be used for designing. Most frequently, however, the
temperature operation is more than occasional and thermal degradation
with time and temperature must be accommodated (Figs. 3.1–3.5).
With machinery adhesives, hot strength determination is more a tradi-
tion than a requirement because they are highly crosslinked (many
attachments on the backbone of the molecules) and therefore behave
as thermoset rather than thermoplastic materials. That is, they do
not melt or revert to a flowable material on being heated. Hot strength
after aging is a more meaningful test for usable properties. Room
temperature strength after heat aging is the best predictor of length
of life, which is further discussed in Sec. 3.

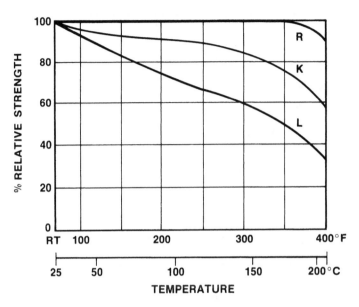

FIGURE 3.1   Hot strength of Grades K, L, and R on steel—Types I and III Newtonian.

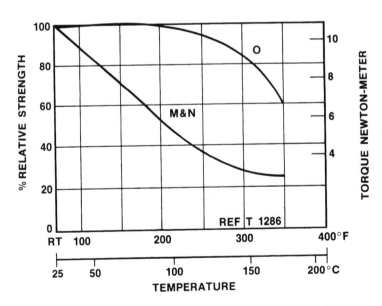

FIGURE 3.2   Hot strength of Formulas M, N, and O on steel—Type II lubricating thixotropic.

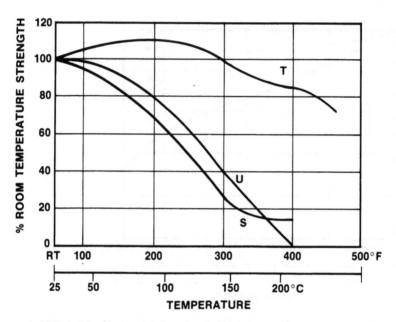

FIGURE 3.3   Hot strength of Formulas S, T, and U on steel—Type IV.

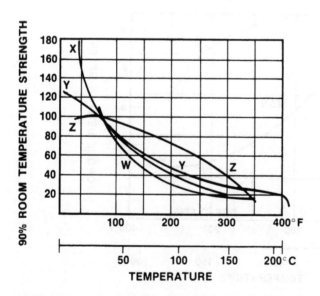

FIGURE 3.4   Hot strength of Formulas W, X, Y, and Z on steel—
Type V.

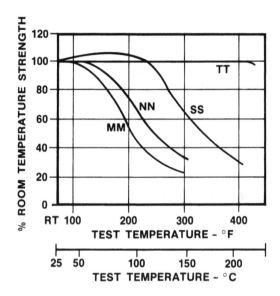

FIGURE 3.5   Hot strength of Formulas MM, NN, SS, and TT on steel—
Type VI.

3. HEAT AGING AND SERVICE LIFE

3.1   Heat Aging Tests

After months or years of exposure to elevated temperature in air,
most organic materials react with oxygen, causing a degradation of
physical properties. Similar degradation occurs under the influence
of moisture, ozone, and many other chemicals. It is most convenient
and practical to test in air without the complication of other factors.
This gives a comparative rating that is highly repeatable.

Machinery adhesives are somewhat more complicated than plastics to
test because they are always tied to a substrate and cure may not be
entirely complete when the degradation test is started.

If the substrate oxidizes at the same time as the adhesive then it
may become as significant to the bond integrity as the adhesive.
This is exactly what happens with adhesively bonded copper and
copper-bearing alloys. The copper forms a loose oxide film under the
adhesive and effectively pries it loose. This occurs only at temperatures over 150°F (66°C) and requires the presence of some moisture.

Fig. 3.6 illustrates pure catalytic degradation of a thermoset plastic.
Some machinery adhesives are not so regular in their aging process.
They always start as monomers or mixtures of monomers and polymers.
The curing process starts after assembly and involves the hooking of
molecules together both end-to-end on the monomer chains and, most

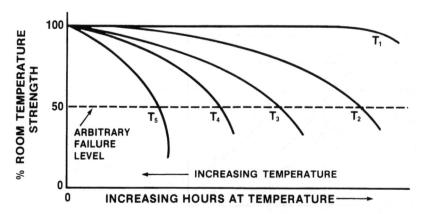

FIGURE 3.6   Heat aging curves for catalytic degradation of a fully cured organic compound.

importantly, at midsections or backbones of the chains. This latter process is called crosslinking. From it result the permanent properties of an irreversible, polymerized, thermoplastic adhesive. Crosslinking occurs rapidly for an hour or so, then continues at a very slow rate somewhat dependent on the temperature. The process is roughly analogous to a bowl of spaghetti which, when freshly cooked and drained of water, flows easily and has no physical shape without restraint of the bowl. If allowed to dry, the flour forms paste and finally glues the long chains together at midpoints until the whole bowlful becomes a relatively homogeneous lump. If this lump is put into the oven to "age" it will get harder and stronger for a while until it gets so dry and crispy that crosslinking bonds start to break and the lump starts to fall apart.

At aging temperature, most machinery adhesives will crosslink more than occurred during the preliminary lower-temperature cure. This means that linking and unlinking are occurring simultaneously, which causes seemingly strange increases in strength and ballooning curves that do not fit the idealized concepts typical of a well-cured plastic (Fig. 3.6). For this reason the curves for S, T, and U in Figs. 3.8– 3.10 are not plotted under 500 hours. Their secondary curing gives rather wild results with little usefulness until they settle into the aging process. The crosslinking effect even reversed the position of the 325 and 300°F curves for Grade U. The Arrhenius technique (Section 3.2) is questionable for this material.

Heat aging of machinery adhesives is usually done in specimen joints of pins and collars, nuts and bolts, or lap-shear strips. Plain steel is the preferred finish. Shear stress is the property observed to deteriorate, and 2000 hours are usually enough to show a trend and to

rate the combination for temperature resistance. Room-temperature
controls are aged for the same time as the heated specimens. The temperature that causes 50% reduction in the room-temperature shear
stress after 2000 hours is called the adhesive temperature limit. Such
a rating is highly arbitrary, for often materials can be used well over
their limits and, conversely, sometimes safe operation is well short of
the limit. The tests are repeatable and provide a way to compare
various formulations. See Figs. 3.7–3.14.

Less emphasis is put onto stressed specimens than is usual with
plastic parts for two reasons. The first is that machinery adhesives
are relatively brittle and are restrained between rigid parts so creep
rupture values will be close to ultimate strength. Second, they are
usually cured completely stress-free in very thin films. The thin films
are usually stressed only in compression during use. Compressively
stressed films are not subject to stress cracking or intergranular
deterioration as would be tensile or tensile shear assemblies.

3.2   Life Extrapolation by Arrhenius Plots

In general, the estimation of service life of any organic material is
predictable on a theoretical basis after certain assumptions are ac-
cepted. We know that, for many nonmetallic materials, the degradation
process can be defined by a single temperature dependent reaction
that follows the Arrhenius equation:

$$k = Ae^{-(Ea/KbT)}$$

where

k = reaction rate	Ea = activation energy
A = frequency factor	Kb = Boltzmann's constant
e = base e = 2.718	T = absolute temperature

For many reactions, the activation energy is considered to be constant over the applicable temperature range.

The accuracy of the Arrhenius method is based on the use of temperature as the sole accelerating means for degrading the adhesive
and the straight-line relationship between temperature and the time
to failure. Underwriters Laboratory Method 746 B requires testing to
be conducted at four or more temperatures. The highest temperature
should last at least 500 hours. The Nuclear Regulatory Commission
will accept data from as short a period as 100 hours.

Any conditions can be imposed in addition to temperature as long
as they are kept constant and only the temperature is changed. For

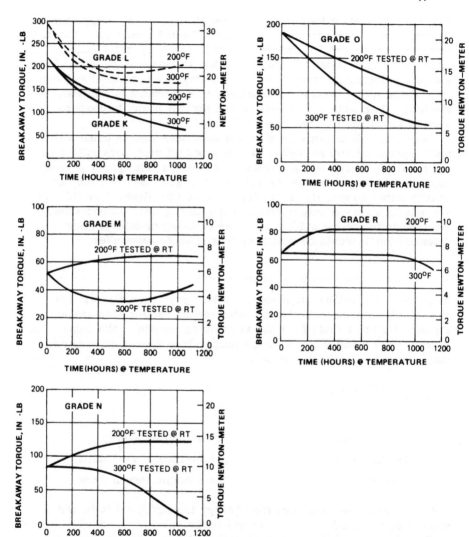

FIGURE 3.7   Heat aging of K, L, M, N, O, and R on steel nuts and bolts.

instance, the tests can be done under liquid water and, as long as the temperature is not so high as to cause the water to change to steam, the extrapolation will give meaningful data. The above assumptions give conservative results for anaerobic machinery adhesives.

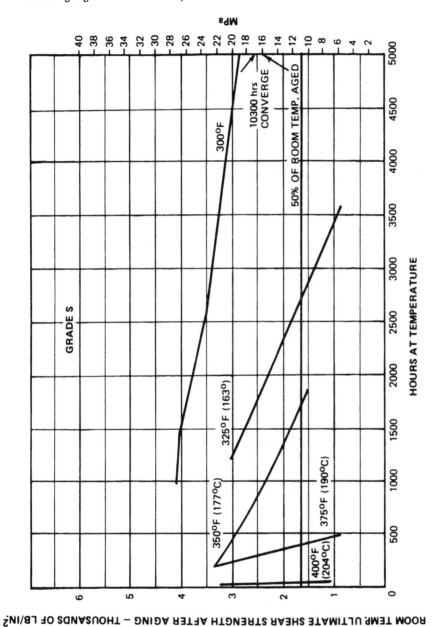

FIGURE 3.8   Heat aging of S on steel pins and collars.

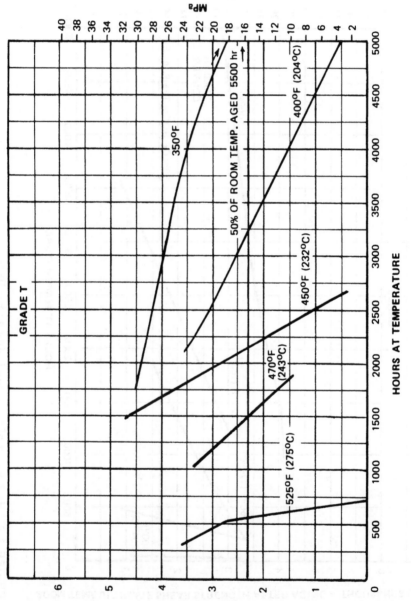

FIGURE 3.9   Heat aging of T on steel pins and collars.

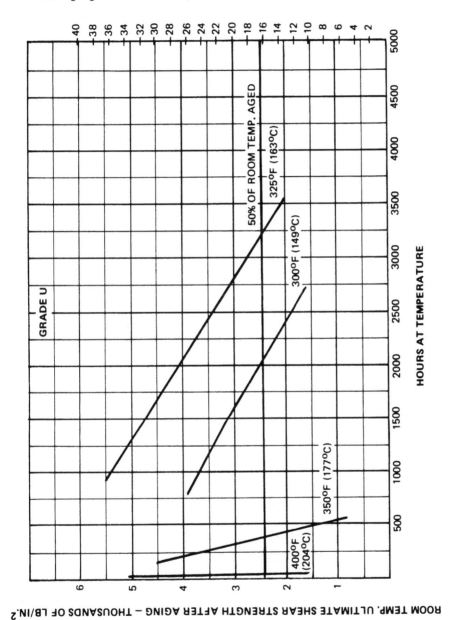

FIGURE 3.10 Heat aging of U on steel pins and collars.

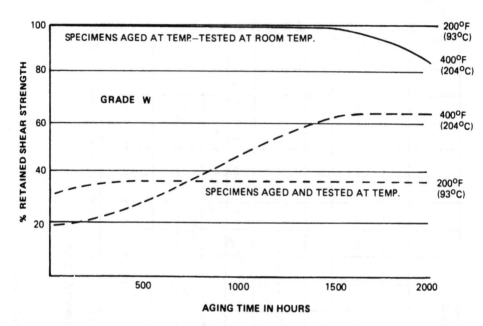

FIGURE 3.11   Heat aging of W on 3/8 in. pipe fittings.

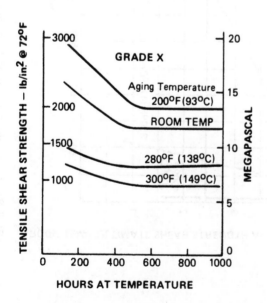

FIGURE 3.12   Heat aging of X on steel lapshears.

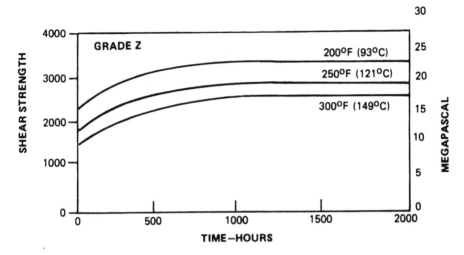

FIGURE 3.13   Heat aging of Z on steel pins and collars.

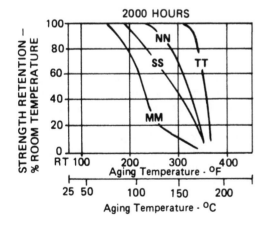

FIGURE 3.14   Heat aging of preapplied materials on zinc phosphated steel nuts and bolts.

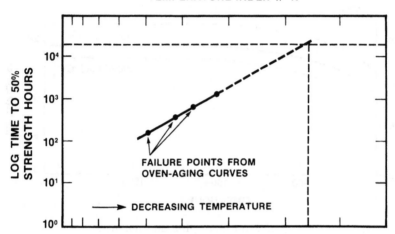

FIGURE 3.15   Arrhenius plot derived from heat-aging curves of
Fig. 3.6 at 50% strength rating.

A plot of the logarithm of the time to "failure" (usually 50% loss of
shear strength for machinery adhesives) versus the reciprocal of the
absolute temperature (°C + 273) should be a straight line.
The data hours are recorded for 50% strength at four temperatures
(Fig. 3.6). The reciprocal of the absolute temperature and the natural
logarithm of hours is computed. These are plotted using a least
squares analysis to give the best fit for a straight line (Fig. 3.15).
The results can be fitted to the following equation:

ln life hours = (Ea/Kb) (1/T) + constant

where the slope of the line is equal to (Ea/Kb) and the constant can
be read from the time intercept. The correlation coefficient for line
fit can also be calculated to establish a confidence level in the data.
Using this technique with the data for Grades S, T, and U from
Figs. 3.8–3.10 gives the prediction-life graph of Fig. 3.16.
For further discussion of Arrhenius plots see IEEE Standard 98-1972
and IEEE Standard 101-1972, which provide useful information on isothermal testing and the analysis of the resulting data.*

---

*IEEE, 345 East 47th Street, New York, NY 10017.

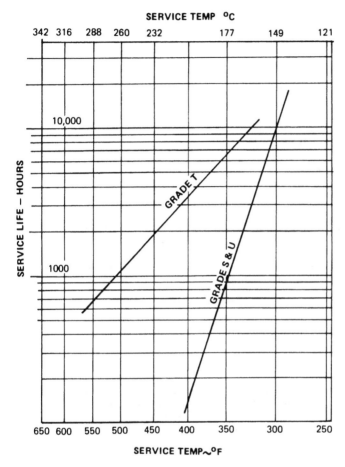

FIGURE 3.16 Arrhenius plot of Grades S, T, and U from curves of Figs. 3.8–3.10.

4. COLD EFFECTS

4.1 Cold Curing

Decreasing the temperature of the system slows the cure of all machinery adhesives. The plot of time vs. temperature shows some formulations curing down to −20°C and others not curing at all (Fig. 3.17). Even though it is possible to make cold-curing formulations, slow cures are not the only problem at reduced temperatures. A more serious problem is the increase in viscosity. For instance, formulations U and Z will cure in 24 hours at 0°F (−18°C) but the viscosity of U is like heavy tar and Z is like chewing gum (Fig. 3.18). It is

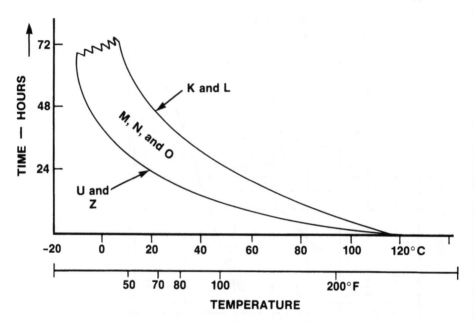

FIGURE 3.17   Time vs. temperature for full cure—formulas U, M,
K, Z, N, L, and O.

necessary to use more fluid formulations and an accelerator or primer.
When this is done, the materials are very usable at 40°F (5°C) and
can be applied and cured as low as 0°F (−18°C). At these low temperatures care must be taken to dry the parts as condensation can
preclude wetting by the adhesive. A wipe with a clean, alcohol-
wetted towel may be necessary.

Drying is especially necessary when a machinery adhesive is used
to augment a sweated fit where one of the parts is chilled to make it
shrink. A deep-frozen part will almost always precipitate ice or frost
when removed to normal atmosphere. It must be wiped with acetone
or alcohol immediately before assembly.

Parts that have started to cure at low temperature and achieved
only a portion of their strength will continue to cure at elevated temperatures without detriment to their final strength. Likewise, materials that have not cured because of low temperature or are even still
in the bottle will not be harmed and will cure normally after they are
rewarmed. For construction applications where parts must be used at
low temperatures and priming is not acceptable, special formulations
are available that cure well and have little viscosity variation with
lowered temperatures.

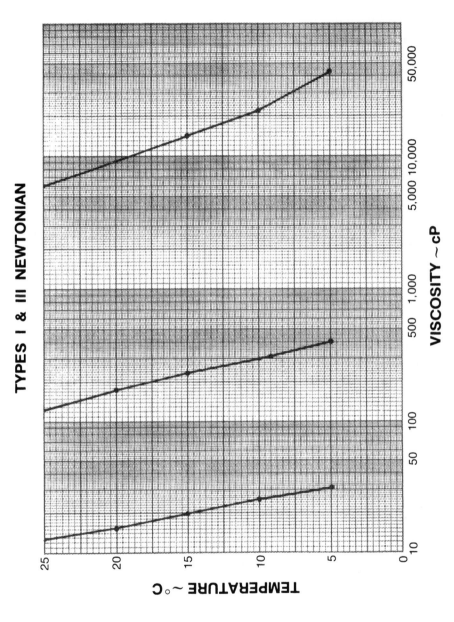

FIGURE 3.18 Typical temperature vs. viscosity—Types I and III Newtonian.

4.2   Cold Strength

At cyrogenic temperatures ($-320°F$, $-160°C$, liquid nitrogen) all
materials become embrittled. Machinery adhesives that are used in
shear and compression are still very effective. No tests have shown
more than a 50% loss. In tests on formulations O and NN the shear
stress was 50 to 90% of the room temperature values. There seems to
be a relationship to the type of substrate; titanium gave the 50% values, whereas Grade 8 steel fasteners into aluminum plate shattered
explosively at very high torques. Undoubtedly, differential thermal
contraction contributed to the latter excessively high values. At less
than cryogenic temperatures ($-110°F$, $-43°C$, dry ice and acetone,
or $0°F$, $-18°C$, food freezer) the shear strength will be equal to or
slightly higher than room-temperature strength. The generalization
may be made that all formulations will perform in shear as well at
$-300°F$ as they will at $300°F$. Flexibility and impact strengths will be
quite low as they are for the materials being joined.

5.   VACUUM AND PRESSURE SEALING

All the machinery adhesives have high compressive strength in thin
films. The way to seal high pressures and space vacuums is to design
them so that compressive stresses do the sealing function and the
joined parts absorb the tensile loads. This is nicely done with standard
pipe fittings. The relatively flexible pipe is screwed into a rigid
fitting. When pressure is applied inside the pipe it tends to expand
in the threaded area, thus tightening the joint (Fig. 3.19). Material

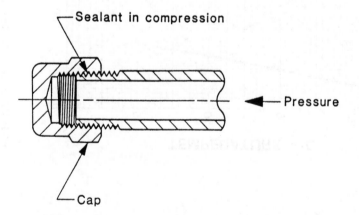

FIGURE 3.19   High-pressure tapered-pipe joint that puts sealant in
compression.

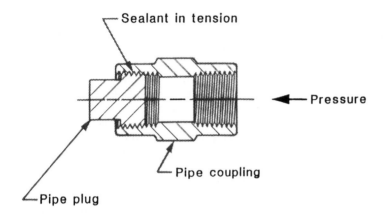

FIGURE 3.20   Low-pressure tapered or straight-pipe joint that can cause progressive failure of sealant in tension.

sealing the helical clearance in the joint will be in compression. Tests have shown that pipes sealed with formula W will seal up to the burst pressure of the pipe which, in actual tests, was 10,000 lb/in. If, however, a solid plug is assembled into a relatively thin pipe or cylinder where adequate tightening is not possible, the expansion of the pipe under pressure is greater than that of the plug and the sealant is placed under tension (Fig. 3.20). If the leak doesn't start immediately, it certainly will on repeated cycles if tensile stresses exceed the fatigue limit. The fatigue limit in tension for all the machinery adhesives is under 1000 lb/in. The reason that these adhesives are so successful as sealants is that their lubricating qualities assist assembly so that relatively high compressive stresses are placed on the joint. These cannot be overcome by internal pressure and the sealant remains in compression.

Vacuums exert much lower pressures and are therefore more tolerant of design looseness. Hermetic sealing may be a consideration but cannot be achieved with any organic materials. All machinery adhesives will pass a few molecules an hour through their molecular lattice so they are not used for true hermetic seals. They are used, however, to seal those small, hard-to-seal molecules of hydrogen, nitrogen, argon, and freon on a routine basis where, in fact, no other organic material will seal as well and metallic seals are impractical.

6. OUTGASSING

Under hard vacuums, smaller than 10^{-8} torr, many plastics, adhesives, and lubricants will allow the escape of certain constituents such as

plasticizers, water, and other lightly linked molecules. This phenomenon was observable in automobiles produced in the '50s and '60s
when plastics were starting to be used extensively for interior upholstery and trim. Not enough attention was paid to outgassing and
the windows quickly fogged on the inside with a tenacious film. This
film can be of serious concern if it is confined within electronic or
switch modules where it forms on optical elements or switch contacts.
For aerospace equipment each material is tested and those with the
least outgassing are preferred in designing of modules.

Two measurements evaluate the outgassing. A very small sample
of cured material is placed in a ceramic boat and baked under high
vacuum. The total weight loss (TWL) is recorded as well as the
weight of volatile, condensable material (VCM) which will plate out
onto a cold metal platen. The limits imposed by the designers at the
National Aeronautics and Space Administration (NASA) are 1% TWL
and 0.1% VCM.

Machinery adhesives that start as relatively volatile liquids do not
normally pass this test and are used extensively as exceptions to the
rule. One rationalization for this is the fact that the tests are done
on free material and not in confined joints where machinery adhesives

TABLE 3.2   Outgassing Data[a]

Material	TWL	VCM	Cure (hours at 25°C)
K	05.24	0.47	48
L	03.79	0.22	72
M	17.71	7.65	24
N	13.74	5.49	24
N	16.43	7.45	72
O	04.43	0.18	72
R	05.19	0.02	24
R	02.97	0.13	48
S	05.21	0.02	24

[a]Variability of results can be accounted for
only by the various specimens used during
cure and the extreme sensitivity of the test
to cure variables. Most data were obtained in
apparatus described in ASTM E 595-77.

are normally used. The thermogravometric instruments are too small
to accept an assembled specimen and are limited to 300 mg per test.
Another way around volatile materials is to postcure and bake out
complete assemblies without their closures. This is simply done with
an atmospheric bake at 212°F (100°C) for 12 hours. This seems to
take care of all materials that might cause trouble by outgassing and
are not harmed by this relative low temperature bake (Table 3.2).

7. FUNGUS RESISTANCE

Conflicting data exist on fungus resistance due to the differences in
test methods and interpretation. Air Force System Command Design
Handbook AFSC DH 1-5 DH 3B2 divides materials into three groups:
    Group I—Fungus inert in all modified states and grades. Under
this class are listed acrylics, glass, silicone-glass fiber, phenolic-
nylon fiber, polypropylene, polysulfone, and silicone resin.
    Group II—Not fungus-resistant in all grades; establish fungus
resistance by test. This class lists acetal resins, cellulose acetate,
epoxy-glass laminates, epoxy resin, lubricants, polyvinyl chloride,
natural and synthetic rubbers, urea formaldehyde, and others.
    Group III—Fungus-susceptible. Listed here are cotton and linen
thread, cork, felt, hair, wood, wool, paper, and leather.
    The machinery adhesives probably fit into Group II since they dif-
fer from formula to formula in their resistance and there is nothing in
their makeup that actually kills fungus. The use of fungicides causes
safety problems as well as odor, color, chemical, and dielectric changes.
The Air Force recommends that whole assemblies be protected in en-
closures or treated with fungus-resistant coatings. For the maximum
resistance when using machinery adhesives choose those without
plasticizers and fillers (e.g., K, L, R). Clean the assembly so that
there is no residual liquid or foreign material remaining on exposed
surfaces, and test.
    Mil Std 810 sect 508.2 has recently superseded Mil Std 5272 as the
test method for fungus resistance. It will pass few organic materials
without fungicidal properties. Machinery adhesives are no exception
and if fungicides are absolutely necessary the manufacturer should be
contacted to establish the possible trade-offs of including a fungicide
in the formulation.

8. CORROSION PREVENTION

What is corrosion? Corrosion is the destruction of a metal by chemical
or electrochemical action. Corrosion of iron-based alloys is called
rusting. The ability to predict corrosion behavior and its effect on

physical properties has made very little progress, although the ability
to explain why corrosion has already occurred has grown.

Every piece of electrically conductive material is a candidate for
corrosion. Any difference in chemical composition, grain size, surface finish, temperature, or grain faces of the same metal can provide
sites for corrosion.

Corrosion develops into several forms. *Pitting* is a localized attack
that generates small pits that may actually penetrate the metal. *Intergranular corrosion* is concentrated at the borders of metal grains.
Stress corrosion and cracking are the result of combined tensile stress
and corrosion acting together. Each proceeds faster with the other
than separately. If the stress is cyclic in nature then *corrosion
fatigue* results.

The electrochemical explanation of corrosion gives the clues to its
prevention. Corrosive action is analogous to an electrical cell. It
works as a tiny battery with an anode, cathode, electrolite, and a
return path that short-circuits the electricity produced. To produce

TABLE 3.3   Galvanic Series in Seawater

1.	Magnesium (most active)	18.	Inconel (active)
2.	Magnesium alloys	19.	Hastelloy B
3.	Zinc	20.	Brasses
4.	Aluminum 1100	21.	Copper
5.	Cadmium	22.	Bronzes
6.	Aluminum 2017	23.	Copper-nickel alloys
7.	Steel	24.	Titanium
8.	Cast iron	25.	Monel
9.	Chromium iron (active)	26.	Silver solder
10.	Nickel cast iron	27.	Nickel (passive)
11.	304 Stainless (active)	28.	Inconel (passive)
12.	316 Stainless (active)	29.	Chromium iron (passive)
13.	Hastelloy C	30.	304 Stainless (passive)
14.	Tin-lead solders	31.	316 Stainless (passive)
15.	Lead	32.	Silver
16.	Tin	33.	Graphite (least active)
17.	Nickel (active)		

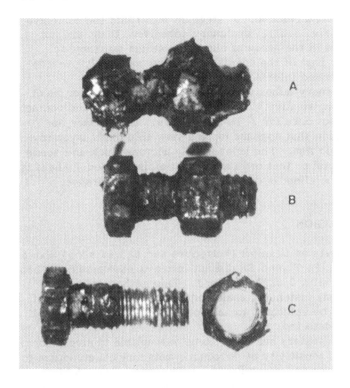

FIGURE 3.21 Six months of corrosion in a salt fog did not reach into the threads that were protected by the adhesive (c).

an electromotive force there must be a chemical difference between the the anode and cathode. The difference can be on a microscopic scale or of a macroscopic size. The macroscopic can be controlled by the selection of contacting materials in accordance with a galvanic series. Unfortunately for predictability, the series varies slightly with the electrolite. Materials should be picked as close together in the series as possible, or at least the most active material (anode) should be designed to sacrifice itself without serious consequences (Table 3.3).

What does all this have to do with machinery adhesives? There are four elements required to produce corrosion:

1. An anode
2. A cathode
3. A reservoir to confine a stagnant electrolite
4. A short circuit path for conducting electricity

Number 3 is where the use of machinery adhesives in close-fitting joints can markedly reduce overall corrosion and completely prevent it in the joint. Without a reservoir of stagnant electrolite the corrosive cell is incomplete (Fig. 3.21). Machinery adhesives, then, are not corrosion inhibitors in the sense of oils and platings but they do eliminate one major part of the corrosive cell and completely prevent corrosion at the bonded interfaces.

The liquid machinery adhesive may have a surface effect on steel and copper when the relative humidity is greater than 50% and contact is maintained without cure for 24 hours or more. The surface develops a discoloration that appears to be a very thin oxide layer covered with an organic film. The layer is only microns thick and seems to passivate the metal so that ordinary oxidation is stopped. There is no evidence that the effect is detrimental except to appearance.

9. STRESS CORROSION

Stress corrosion tests on titanium (both pure and 6A1 4V alloy) showed that formulas N, O, P, T, and SS did not cause stress corrosion when the test method in Mil N 25027D was used. This method calls for stressed (90,000 psi) fasteners, sealed and locked with an excess of liquid adhesive, to be subjected to salt fog for a week. No effect of any kind could be detected.

High-strength fasteners are likely to be susceptible to stress corrosion because of the sensitivity of martinsitic metallurgical structure to the incursion of water and other fluids. Under corrosive conditions care should be taken to avoid corrosion by plating the fasteners properly and avoiding stresses over 100,000 psi on high carbon steels. There is no evidence that the use of machinery adhesives either chemically accelerates or retards stress corrosion except that the corrosion reservoir in the threads is physically eliminated.

10. EXPLOSIVE COMPATIBILITY

Reagents in explosive materials are very active chemical combiners and some of the compounds used in anaerobic adhesives will react with them. Special formulations have been developed which are compatible with selected explosives and propellants. Every combination must be separately tested. It is not possible to list compatibility in a compilation such as this. Consult a manufacturer for details of compatibility if any crosslinking material is to come in contact with an explosive.

Chapter 4
Application Methods
and Safety

1. INTRODUCTION

Much emphasis justifiably is placed on the selection of the proper
machinery adhesive. But proper application is equally important.
Responsible specialty adhesive companies emphasize a systems approach. It includes adhesive selection plus four application considerations:

1. Cleanliness of parts as shown by the adhesive's ability to wet
   them
2. Proper application
3. Quality assurance
4. Safety

   1. Cleanliness of the parts and the ability of the adhesive to wet
the surfaces is of primary importance. A simple water-break test can
often show whether the parts are clean or need cleaning. Just place
a drop of water onto the surface and observe whether it spreads or
stays beaded up as it would on a waxed surface. Next, the same test
can be tried with the selected adhesive. If the water does not wet and
the adhesive does then you can expect good performance but not as
good as you might get with clean parts. Formulations M, N, and O
are designed specifically to cut through light oil films and give satisfactory results. There will, however, be a loss of adhesive properties
and, with gross contamination, some softening of the crosslinked
material. In any case the adhesive must be able to wet the parts since
complete fill of closely fitted parts always depends on wetting and
capillary action to complete the fill and exclude air bubbles and
pockets.

The sources of contamination are always changing so an adhesion problem once solved may not stay that way. Changes in washing solutions and rust preventatives are fairly obvious sources of wetting and cure failures. Who would expect that the polish put on the floor by the cleaning people could somehow contaminate the top layer of parts previously degreased and ready for assembly? How about that silicone spray furniture polish, the open steam humidifier that spread boiler treatment chemicals into the air ducts, the plumber using acid core solder, or the window washing solution from an aerosol spray? Fortunately the performance of machinery adhesives is not super-sensitive to contamination. The major reason for failure of a joint, in the author's experience, has been the absence of adhesive because of lack of wetting or incomplete application.

2. Proper application is the next part of the complete system consideration. You cannot haphazardly apply adhesive without risking bond integrity and wasting material. The correct amount of adhesive must be applied in the right place even with simple rolling or brushing applications.

Proper application begins all the way back at the design stage, for closely fitted parts vary considerably in "fill" volume, that is, the amount of adhesive needed to fill the joint. A Class 2, 3/8-16 thread has a volume variation within its tolerance of 8:1. Accordingly, you should provide a reservoir or space for excess adhesive for all parts with less than maximum clearance.

You can easily remove excess material where required after the joint has cured. This is done in subsequent washes or vapor degreasing. Be sure to include it in your plans.

When designing a shaft for adhesive bonding, include a starting chamfer or recess on the bore both for ease of assembly and as a reservoir for filling the variable space. A shaft with a 0.001 in. tolerance and 0.001 in. fit in a bore with a 0.002 in. tolerance has a 3:1 variation in fill volume even when all parts are within tolerance.

The quantities suggested in Chap. 2 are based on the standard tolerances of Class 2 fasteners and are liberal enough to accommodate most production variables.

3. Inspection of assembled parts is an important manufacturing function on any device. There are several ways to be sure that material is in place. Low-level warning devices are available on most applicator feed reservoirs so that you are sure of constant supply. An automatic machine can inspect just after application and prior to assembly by means of a sensitive photodetector. All machinery adhesives are fluorescent under a "black" light such as those used to inspect minerals. The detector can pick up the glow before assembly or, if an excess is used, even after assembly.

Functional destructive or nondestructive tests are also used. In the nondestructive category is the push-out force test on shafts. It is often done on the assembly machine at a carefully selected force that will not move good parts but detects lack of fill or cure.

Threaded parts are often satisfactorily tested at a torque 10% higher than the initial assembly torque and in the same direction. If any movement of the fastener occurs the assembly is defective.

You can statistically function test each production lot of dry adhesives, grades MM, NN, SS, and TT. This is done by assembling a nut and bolt and checking the on-torque during assembly, allowing time for cure, and checking the break-away torque (untightened bolt). Assembly on-torque should be below the manufacturer's maximum. Excessive torque indicates that the shelf life has been exceeded. The cured break torque should be above a preestablished minimum to assure that the ultimate performance is still available. Visual inspection can determine overall lot consistency of coating.

4. Improved safety is a direct benefit of automating adhesive application and assembly. You can provide good ventilation and separate operators from moving parts, machinery, and adhesives. Most applicators are loaded by placing the shipping container, bottle, or cartridge directly into a pressure container or other emptying device.

2. APPLICATION OF LIQUIDS

2.1 Low-Cost Application Aids

It does not make sense to start with a complicated system of application if a simple inexpensive technique can start one up the learning curve before large commitments are made. There are simple applicators and techniques that have been proven all over the world by equipment manufacturers and repair service operations. Some of the devices suggested are especially made for adhesives; others are in general use as, for instance, an envelope moistener.

It is not suggested that these devices and techniques are in any way a substitute for highly controlled automated equipment. You are advised, however, to put parts together carefully by hand in fairly substantial quantities before opting for expensive automation. The only shortcut that works is to use specialty adhesive engineering services. Even they will hand-assemble many parts, simulating proposed methods before fixing automation parameters.

For limited production and maintenance operations the application aids shown in Figs. 4.1–4.25 may be all that you will ever need. Each costs less than $30.

Drop Applicators

Use with Grade: K, L, M, N, O, R, S, T, U.

Where available: Local Loctite Distributor, Item #10976.

Benefits: Dispenses controlled drop size repeatedly in range
                    from 0.01 to 0.04 ml.

How to use: Replace the cap on a 250 ml bottle with the hand-
                    gun. Prime by rotating the knob. Adjust the
                    quantity using the screw in the trigger. Nozzle
                    may be fitted with standard needles for pinpoint
                    control.

FIGURE 4.1   Bottle top handgun.

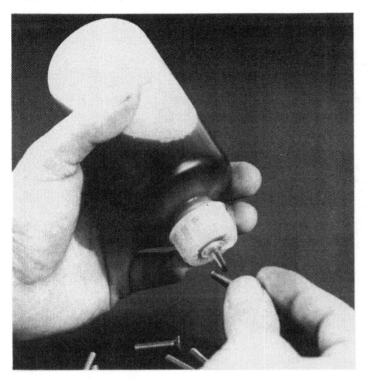

Drop Applicators

Use with Grade: K, L, M, N, O, R, S, T, U.

Where available: Cole-Palmer Instrument Co., 7425 N. Oak Park
                    Ave., Chicago, IL 60648, Item #6086-50.

Benefits: Applies drop with squeeze bottle.

How to use: Pour liquid into the bottle. Turn over and squeeze
                    out the desired amount.

FIGURE 4.2   Drop dispenser bottle.

Drop Applicators

Use with Grade: K, L, M, N, O, R, S, T, U.

Where available: Drug- and department stores.

Benefits: Low-cost, efficient spreading of adhesives and ac-
                   celerators.

How to use: Dip the Q-Tip into a small container of liquid. Do
                   not return used material to storage container.

FIGURE 4.3   Cotton swab (Q-Tip).

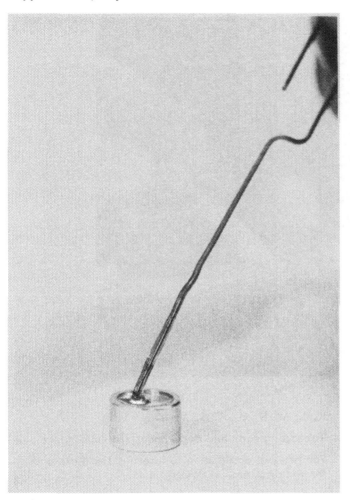

Drop Applicators

Use with Grade: K, L, M, N, O, R, S, T, U.

Where available: Drug-, department, stationery, and variety stores.

Benefits: Low-cost drop transfer.

How to use: Use a #1 paper clip for liquids when only a very small amount of product is needed. Bend clip into shape shown; hooked end will deposit a drop 1/8 in. in diameter and the straight end 1/16 in. in diameter.

FIGURE 4.4 Paper clip.

Drop Applicators

Use with Grade: K, L, M, N, O, R, S, T, U.

Where available: Tobacco, drug- and department stores.

Benefits: For product application to internal diameters and for wicking into assembled parts.

How to use: Bend pipe cleaner in half and use like a miniature paintbrush for drops or a film on small surfaces.

FIGURE 4.5   Pipe cleaner.

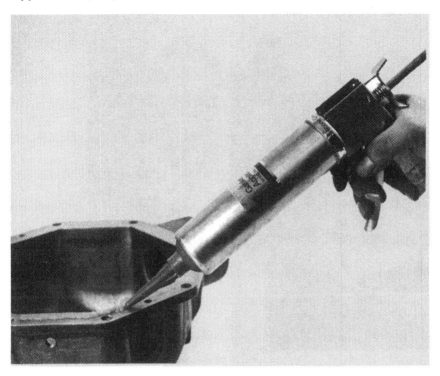

Bead Applicators

Use with Grade: W, X, Y, Z.

Where available: Authorized adhesive distributor.

Benefits: Provides positive cutoff without dribble when apply-
                     ing beads of material.

How to use: Load 160 ml cartridge into gun and squeeze trigger.

FIGURE 4.6   Hand cartridge gun.

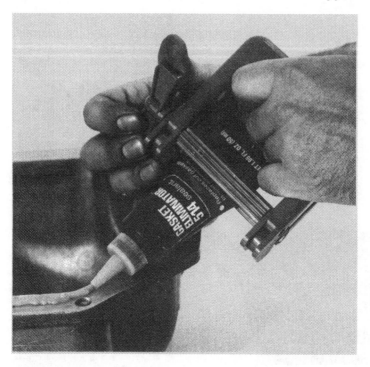

Bead Applicators

Use with Grade: W, X, Y, Z.

Where available: Authorized adhesive distributor.

Benefits: Completely empties tubed products, eliminating
                     waste.

How to use: Clamp on the end of plastic tube. Turn handle to
                     dispense material.

FIGURE 4.7   Tube wringer.

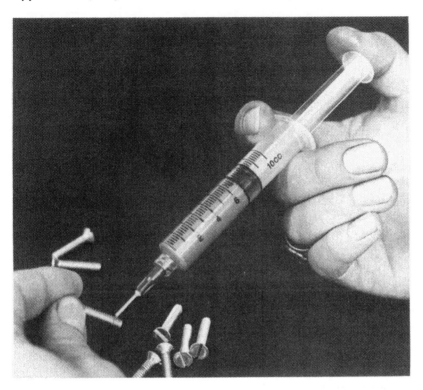

Bead Applicators

Use with Grade: K, L, M, N, O, T, U.

Where available: Becton Dickinson & Co., Cornelia St., Rutherford, NJ 07070; medical supply houses.

Benefits: For bead or drop applications.

How to use: Fill by inserting syringe tip into liquid container. Pull plunger back slowly. Not recommended for low viscosity.

FIGURE 4.8 Disposable syringe, 10 ml.

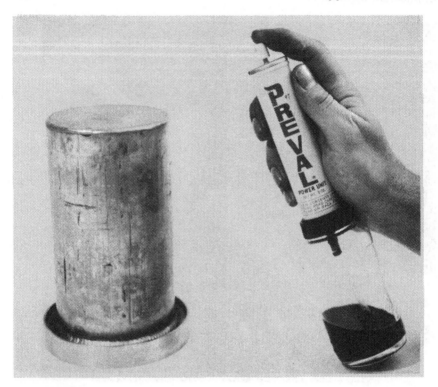

Bead Applicators

Use with Grade: R.

Where available: Hardware stores, hobby shops, auto paint stores.

Benefits: Sprays low-viscosity products.

How to use: Pour sealant into Preval bottle. Hold sprayer 12 in.
                     from surface to be sprayed for best spray pattern.
                     Will spray up to 500 ml of Grade R per propellant
                     refill. Requires good ventilation during use.

FIGURE 4.9   Preval bottle top spray.

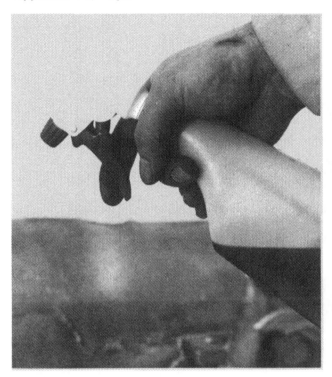

Bead Applicators

Use with Grade: R.

Where available: Authorized local Loctite industrial distributor,
                   Loctite Item #11587.

Benefits: Sprays low-viscosity products. Pumps high-viscos-
                   ity products.

How to use: Pour product into the plastic bottle. Pump trigger
                   to spray. Spray pattern can be adjusted by ro-
                   tating nozzle. Provides a wet spray so venting is
                   not required.

FIGURE 4.10   Pump spray bottle.

Bead Applicators

Use with Grade: R, S.

Where available: Sterling Plastics, Mountainside, NJ 07092, Sterling
                    Item #2; stationery suppliers; drug stores.

Benefits: Applies product to wide surface area by wiping.

How to use: Pour sealant into the bottle. Turn bottle over and
                    squeeze while wiping surface. With Grade S the
                    hole in the cap should be enlarged to 1/16 in. in
                    diameter. Do not wipe primed surfaces.

FIGURE 4.11   Squeeze bottle moistener.

Bead Applicators

Use with Grade: R, S.

Where available: Stationery suppliers, department stores, drug
                    stores.

Benefits: Sponge applicator fits 50 ml bottle of Grades R and
                    S. Applies product to corners.

How to use: Remove sponge appliator from the moistener and the
                    plug from the 50 ml bottle. Insert the sponge ap-
                    plicator, invert the bottle, and squeeze. With
                    Grade S the hole in the cap should be enlarged to
                    1/16 in. in diameter. Do not wipe primed surfaces.

FIGURE 4.12   Envelope moistener.

Dip and Roll Coating Applicators

Use with Grade: K, M, N, O, R, S, U.

Where available: A. E. Aubin Co., 134 Pine St., Manchester, CT
                     06040, (203) 643-0321.

Benefits: Dip application for hand-held parts.

How to use: Remove bottle cap and plug of 50 ml bottle. Insert
                     upright bottle into the constant level reservoir.
                     Turn assembly over slowly allowing product to flow
                     to dip hole. Caution: Reservoir may overflow with
                     barometric changes.

FIGURE 4.13   Constant level reservoir.

Dip and Roll Coating Applicators

Use with Grade: K, L, M, N, O, R, S, T, U, Primers N and T.

Where available: Stationery suppliers.

Benefits: Quick-dip applicator for screws.

How to use: Saturate sponge with product. Push part into sponge to desired coating level.

FIGURE 4.14   Sponge cup.

Dip and Roll Coating Applicators

Use with Grade: K, L, M, N, O, R, S, T, U, Primers N and T.

Where available: Variety stores, drugstores, department stores, stationery suppliers.

Benefits: Good for application to outside diameter of cylindrical parts.

How to use: Saturate pad with product. Roll part over pad. Cover when not in use—especially with activators.

FIGURE 4.15   Carter's felt stamp pad.

Dip and Roll Coating Applicators

Use with Grade: K, L, M, N, O, R, S, T, U.

Where available: Variety stores, drugstores, department stores, stationery suppliers.

Benefits: Quick-dip applicator for small screws.

How to use: Saturate pad with product. Dip parts up to 1/4 in. wide into sponge or roll part over sponge.

FIGURE 4.16   Carter's foam stamp pad.

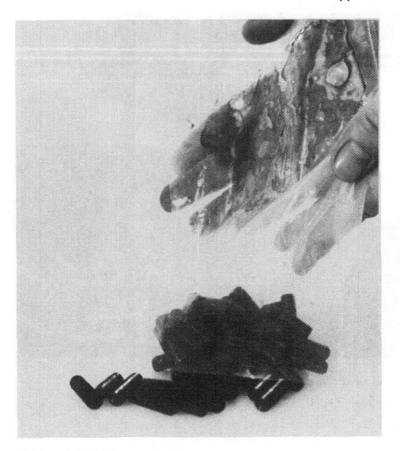

Dip and Roll Coating Applicators

Use with Grade: K, L, M, N, O, R, S, T, U.

Where available: Local adhesive industrial distributor.

Benefits: Disposal tumbler for 100% coverage of batches of parts.

How to use: Put parts in bag. Consult reference chart in Chap. 4 and pour correct quantity of sealant into bag. Close bag and shake until coverage is uniform.

FIGURE 4.17   Polyethylene bag.

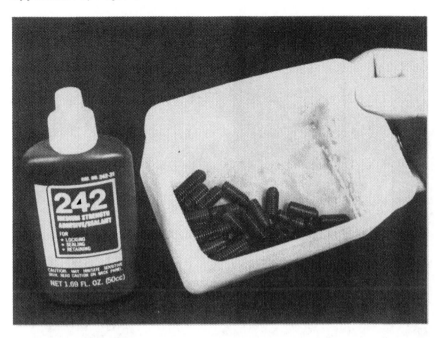

Dip and Roll Coating Applicators

Use with Grade: K, L, M, N, O, R, S, T, U.

Where available: Discount and variety stores.

Benefits: Low-cost tumbler for 100% coverage of batches of parts.

How to use: Put parts in tub. Pour in material. Shake tub until uniform coverage is obtained.

FIGURE 4.18   Polyethylene food box.

Dip and Roll Coating Applicators

Use with Grade:	K, M, N, R, S, U.
Where available:	IDL Mfg. & Sales Corp., Carlstadt, NJ 07072, (201) 933-5774; stationery suppliers.
Benefits:	Provides controlled film thickness for application to outside diameter of cylindrical parts.
How to use:	Fill reservoir with product. Turn part against roller. Recommended for products up to a viscosity of 1500 cP. Replace roller axle with a wood or plastic one to minimize compatibility problems.

FIGURE 4.19 Porcelain stamp and label moistener.

Miscellaneous Applicators

Use with Grade: L, T, X, Y, Z.

Where available: Floor-covering stores; department stores; Sears, Roebuck & Co., Catalog Item #37A5013NPH.

Benefits: Provides transfer coating of large surfaces.

How to use: Saturate a section of carpet with product. Press part into the carpet. Lift straight up and check coverage.

FIGURE 4.20 Indoor—outdoor carpet.

Miscellaneous Applicators

Use with Grade: K, L, M, N, O, R, S, T, U.

Where available: Local Loctite industrial distributor, Loctite Item #11730.

Benefits: Convenient dip cup prevents contamination of bulk of product.

How to use: Remove top cup and pour product into the bottle. Replace cup and squeeze bottle until sufficient material is in the cup. Dip Q-Tip, paper clip, brush, pipe cleaner, or the part into the cup.

FIGURE 4.21   Measuring dip cup.

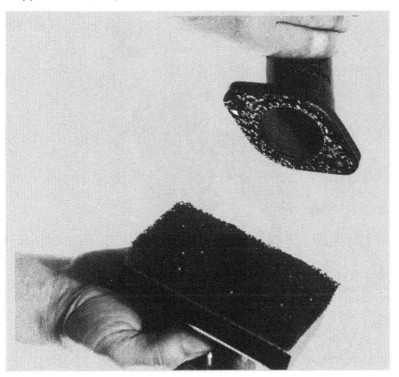

Miscellaneous Applicators

Use with Grade: X, Y, Z.

Where available: Rogers Foam Corp., 114 Central Street, Somer-
ville, MA 02134, (617) 623-3010.

Benefits: Inexpensive transfer technique, especially useful
for irregular surfaces. Transfers large quantity.

How to use: Saturate a section of foam with product. Press
part into foam or attach foam to a piece of wood and
touch part.

FIGURE 4.22   Foam pad transfer.

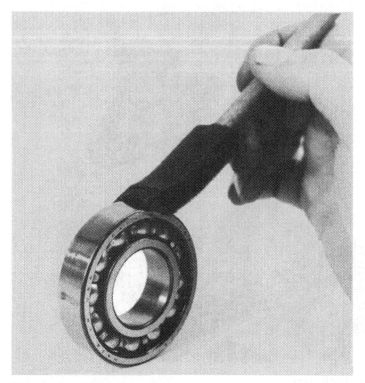

Miscellaneous Applicators

Use with Grade: K, L, M, N, O, R, S, T, U, Z.

Where available: Hardware stores, department stores, paint stores, variety stores.

Benefits: Disposable appliator to provide uniform coat of material.

How to use: Use as paintbrush in conjunction with the dip cup.

FIGURE 4.23   Plastic foam paintbrush.

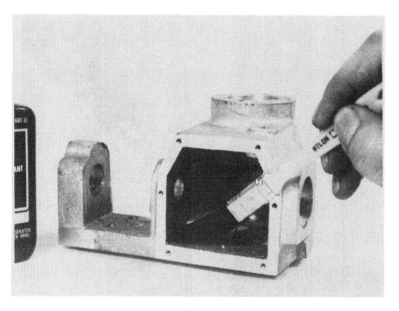

Miscellaneous Applicators

Use with Grade: K, L, M, N, O, R, S, T, U.

Where available: Hardware stores, department stores, paint stores.

Benefits: Disposable applicator for surface coating.

How to use: Use as paintbrush in conjunction with dip cup. Brushes with bristles stapled into handle are recommended.

FIGURE 4.24   Nylon bristle paintbrush.

Miscellaneous Applicators

Use with Grade: X, Y.

Where available: Sears, Roebuck & Co., Catalog Item #11-469; paint
                   stores; variety stores.

Benefits: Easy surface coating of wide flanges.

How to use: Using short nap (1/8 in.) roller, saturate roller,
                   apply material to wide flanges. For narrow flanges
                   (1/4 in. or less) use a piece of indoor—outdoor
                   carpet.

FIGURE 4.25   Paint roller.

2.2 Systems and Automation

Every successful assembly requires that the variables be controlled. Adhesive assembly is no exception. Although manual operations can be skillfully regulated, the best control is obtained with some degree of automation. Here is how automation and a systems approach can control some of the variables of adhesive assembly.

1. Precise handling of the parts controls
   a. The positioning for assembly accuracy
   b. The place of application
   c. The rate of transfer of heat
   d. The positioning for automatic inspection
2. Placement of the exact quantity in exactly the right place controls
   a. Migration and wetted area
   b. Volume filled and reservoir position
   c. Air bubbles and voids by preventing their formation
   d. Cost of material by eliminating waste
3. Precise machine cycle time controls
   a. Fixture and curing time
   b. Productivity and labor cost
   c. Application and wetting time
4. Ovens and other environmental devices control
   a. Curing temperature
   b. Material viscosity
   c. Humidity
   d. Cleanliness of the environment and parts
5. Chemical cleaners, vapor degreasers, and sandblast equipment control
   a. Part cleanliness
   b. Activity of the surface to promote cure
   c. Corrosion before assembly
   d. Surface finish

How does one go about automating an adhesive assembly operation? The best advice is the same as with any assembly operation. Follow logical small steps. Don't automate a sloppy or inefficient current operation just because an adhesive has been added. Achieve the best hand operation possible by adding one automation element at a time. The first element might be a standard applicator to control the liquid variables of quantity, position, voids, and speed.

The automation of parts requires considerable experience and machine know-how beyond the scope of this handbook. However, there are two warnings that may benefit anyone who is considering automation to control costs and quality. The first is to observe the manual operation closely. Look for operator discretionary actions. These are usually evidenced by a box of rejected parts or assemblies. A machine cannot discriminate or use discretion unless it is specifically designed to do so. Four nuts with tight threads out of 4000 assembled a day can completely bog down or damage a machine but are scarcely noticeable to a human assembler who throws them aside.

The second piece of advice is to avoid the trap of average numbers. As every part has a tolerance (sometimes two, one on the drawing and another different one produced in the shop) every adhesive has a variance in cure speed, strength, viscosity, etc. Some of these variables are interactive, such as gap and cure speed. Careful studies should be made of actual parts and the extremes of their ranges. The machinery should be designed to handle the *extremes*, not the average of "typical" values. Specialty adhesive manufacturers with equipment capability can help avoid some of the common traps, but the part manufacturer knows his own parts best. A joint effort with the free exchange of information works best of all. (See Table 4.1.)

Automatic Dispensing Pumps

Automatic dispensers generally consist of a pumping system, a valve, and an application head. There are two pumping systems commonly used: a pressurized reservoir and a positive displacement piston.

The pressurized reservoir is used with a timer that actuates a valve; hence it is called a pressure-time system (Fig. 4.26). Discharge quantity is controlled by a combination of pressure variation, valve opening stroke, and time. It is capable, therefore, of generating either spots or continuous lines. The amount of liquid dispensed through a #16 gauge needle (0.047 in. or 1.1 mm inside diameter) versus the reservoir pressure is plotted in Fig. 4.27.

The positive displacement pump system varies its output by means of a variable piston stroke. Discharge can be made continuous only by pumping into an accumulator (sometimes simulated by a long, flexible discharge line). Normally the piston pump is used to give very accurate spots of material (Fig. 4.28).

It should be understood that all of the equipment described is especially designed to be compatible with the destabilization caused by the exclusion of air from anaerobic materials. Normal plastic or metallic valves and reservoirs are entirely unsuitable.

The applicator heads are used to adapt to the size and shape of the parts and the type of spot or bead desired. The head sometimes includes the control valve, as it is desirable to have the valve close to

TABLE 4.1 Application Parameters of Various Methods

Application Method	Rate	Quantity Range	Viscosity Range Pa.sec	Grades	Typical Application
• Spot or Drop					
Hand	10/min.	0.01-0.04ml	0.001-10	K, L, M, N, O ⎫	Hand assembly
Automatic	100/min.	0.005-0.01ml	0.001-50	R, S, T, U, ⎭	Nut or bolt treater
• Extrude-Pressure Time –Transfer Pump	200in/min. / 9 in./sec.	1/16 to 1/8" bead / 1/32" to 1/16" bead	0.01-8	K, L, M, N, O, T, U / W, X, Y, Z	Flanges, Pipe threads / Flanges with Robot
• Roller -Hand	10 parts/min.	0.001-0.005" th'k	5-5000	L, T, W, X, Y	Flat flanges
• Tumble	5 barrels/hr.	0.01-.05ml/part	0.5-500	K, L, W, & spec.	Pipe plugs & screws
• Screen-Manual / Automatic	4/min. / 30/min.	0.001-0.015in th'k / 0.001-0.015in th'k	>50 / >50 ⎫⎭	W, X, Y, Z.	Flanges
• Screen Transfer pad	2 parts/min.	0.001-0.004in. th'k	>50	W, X, Y, Z	Interrupted flanges
• Spray-Manual -Automatic	5 parts/min. / 20 parts/min. ⎫⎭	0.01-1 ml	0.001-0.1	R, S	{ Castings. Welds. / Transformers
• Rotary Sprayer	6 cycles/min.	0.01-0.1ml	1-50	K, L, M, N, O, S, T, U	{ Core holes, axle tubes / { Sleeves
• Impregnation Vacuum-pressure	3-5 baskets/hr	0.1-50ml/part	0.001-0.1	R, S special	Powdered metal / Die castings

Note: Some grades require special processing for the best results. Extruded materials should be screened and filtered to prevent gaps in the bead. Tumbling and impregnating materials may need special stabilizing to give satisfactory processing life.

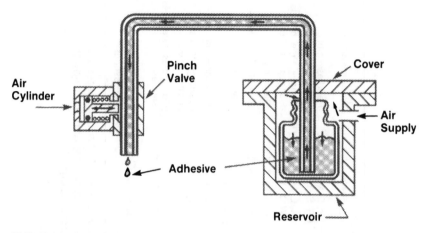

FIGURE 4.26   Schematic of a pressure-time system.

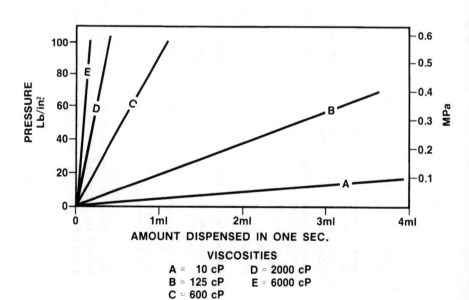

FIGURE 4.27   Pressure-time system liquid-flow rates.

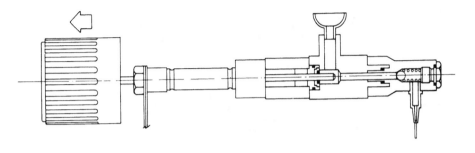

(a)

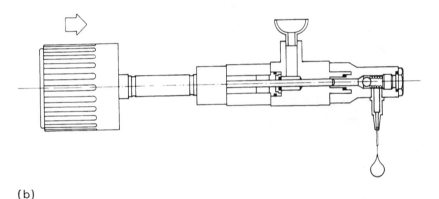

(b)

FIGURE 4.28   Piston pump schematic.   (a) Suction; (b) discharge.

the discharge orifice to prevent after-dribble. As a further feature,
most modern valves control after-dribble with a suck-back feature that
gives a sharp cutoff of each dispense cycle. (See Figs. 4.29–4.33.)

Complete systems can be put together with these dispensing pumps,
heads, and parts handling equipment. Many millions of dollars worth
of such systems are currently working. The applications illustrated
in Figs. 4.34–4.36 are typical.

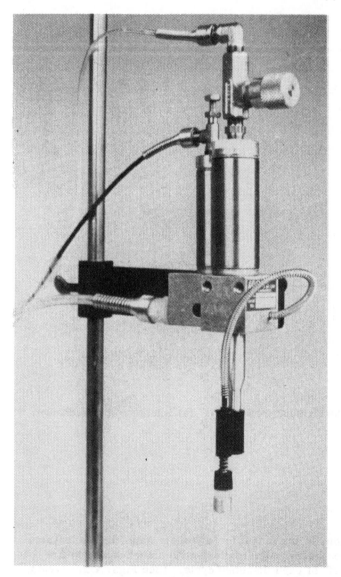

FIGURE 4.29   Advancing head with integral valve.

FIGURE 4.30 Touch applicator with integral valve and actuator.

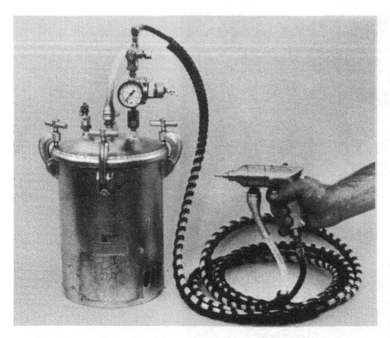

FIGURE 4.31   Manual spray gun with integral valve and actuator and reservoir.

Screening and Stenciling

The anaerobic nature of these materials makes them uniquely suitable among adhesives for screening and stenciling. Photo-resist screens like those used in ink printing can be used to produce intricate patterns on flat parts. The screens are made from nylon, polyester, or stainless steel. Mesh sizes are chosen on the basis of the amount of adhesive to be deposited (Fig. 4.37). Likewise, stencils made of stainless steel or Teflon*-coated steel are able to control quantity and position very accurately. Quantity is determined by the volume of the holes in the screen or stencil. Deposits of 0.001 to 0.015 in. thick (0.04 to 0.4 mm) can be made in a few seconds. The best materials to use for screen application are those that remeld after printing so that air bubbles are not trapped. Grades X and Y are often used for screening (Figs. 4.38 and 4.39).

---

*Registered trademark, E. I. duPont de Nemours & Co.

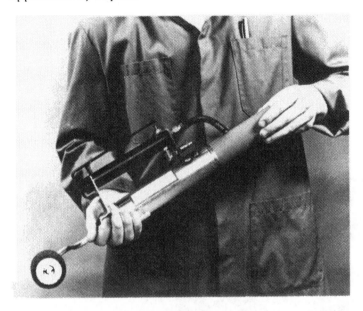

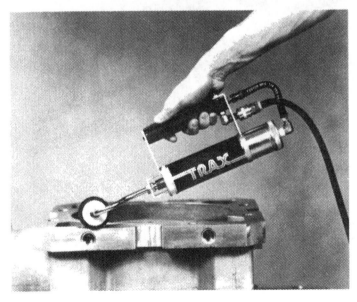

FIGURE 4.32 Automatically fed hand roller; (top) loading of cartridge.

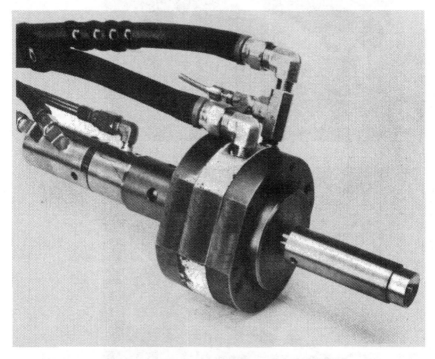

FIGURE 4.33   Rotary sprayer for coating of internal bores. Discharge is through the rotating spindle on the right and into the
rotating end cup, from which it is flung onto the inside of a hole.

FIGURE 4.34 An application system for automating the application of adhesive for eight screws in an automatic transmission cover.

FIGURE 4.35   Roto-Spray of core holes to seal cup plugs in automobile heads.

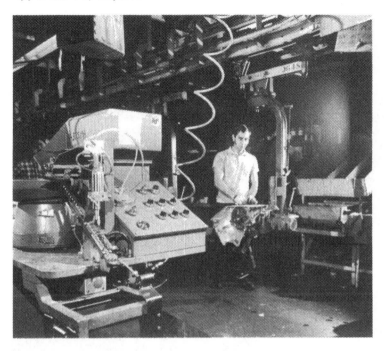

FIGURE 4.36 Automatic nut treater for transmission linkage nuts.

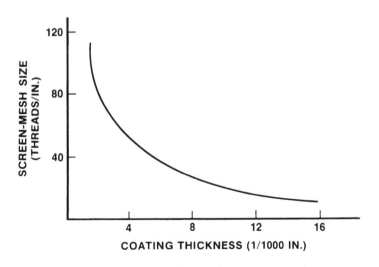

FIGURE 4.37 Thickness of sealant for various screen-mesh sizes.

FIGURE 4.38   Semiautomatic screen printing.

Tracing

Tracing by means of programmed machines is becoming more and more practical as simpler and cheaper machines become available. A pumping system is actuated by the tracer with a dispensing head and valve mounted on the tracing head. An intricate pattern can be traced in a matter of seconds. The rate of application is determined by the extrusion rate and the speed of the tracer. Common tracing speeds range from 3 to 9 in./sec (75–230 mm/sec). (See Fig. 4.40.)

Vacuum Impregnating

Most castings and welds and all powder metal parts have inherent microporosity, which allows seepage of liquids and gases. The microporosity (0.005 in. or 0.1 mm diameter) does not affect structural integrity but does require either extra-thick parts to prevent seepage or, for best cost and weight effectiveness, thin parts plus anaerobic resin impregnation (Chap. 7, Sec. 12).

FIGURE 4.39 Screen transfer printing where the print is made on a
plate, which then transfers it onto a recessed flange of the part.
The flat screen is unable to print directly onto the recess because the
seal area is recessed. Parts being coated are shown on the rotary
table.

Impregnation of parts by immersion in an anaerobic liquid is usually
automated on a batch basis. Since most porosity consists of blind
holes filled with air, the parts must be evacuated in a vacuum vessel.
The evacuation is done either while the parts are under the liquid
or before immersion. In either case, the parts are evacuated, immersed, and pressurized to produce deep penetration. A spin to
remove excess material prepares them for a wash-and-cure dip. The
parts come out very clean with no residue, either cured or uncured,
to be removed. Even holes as small as 0.06 in. (1.5 mm) can be kept
clean with proper materials and processing. The maximum hole size
filled and cured on a single cycle is less than 0.01 in. (0.3 mm).

(a)

FIGURE 4.40   Flange tracing with extruded sealant.   (a) Engine block in place.   (b) Closeup of split bearing bore being sealed.

This is fortunate because larger holes may affect structural integrity. The process is limited (automatically by its anaerobic nature) to sealing only microporosity.

Large parts are often sprayed or brushed with sealants R or S. Large production quantities may justify special equipment and resins to control function and costs.   An anaerobic impregnation equipment specialist should be consulted.   (See Figs. 4.41−4.43.)

(b)

FIGURE 4.40   (Continued)

(a)   (b)

FIGURE 4.41   Carburetor cutaway showing microporosity in the drilled hole and macroporosity on the cut face.

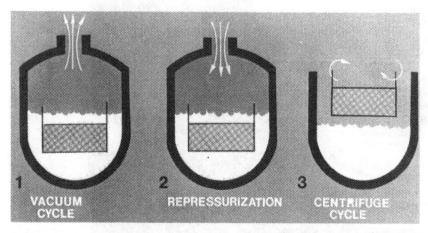

(a)

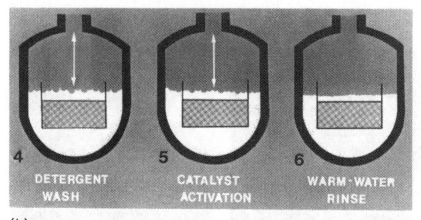

(b)

FIGURE 4.42   Schematic of an impregnation system.

FIGURE 4.43 Industrial impregnation system for die-cast carburetors includes parts-handling basket, spinner, refrigeration and air supply to stabilize the resin, and a vacuum pump with controls.

3. APPLICATION TIPS

In the early days of machinery adhesives, there were two problems that caused the complaint "the stuff just doesn't work." The first problem was caused by the capricious nature of the cure system, which was ultrasensitive to dirt. This problem has been largely solved with newer materials that are tolerant of most common shop dirt and oils. Parts do not require thorough cleaning or activating in most cases.

The complaint that still remains is that of low performance when investigation often reveals little or no material in the joint. It is worth looking at the reasons for lack of joint fill:

1. Material is too thick to wick into voids left by the assembly technique used.
2. Material is too thin to stay in the gap. It runs right through.
3. The assembly technique wipes the parts clean.
4. The assembly of parts creates trapped pockets of air.
5. The material is too fast to allow wicking and flowing to take place before it gels.
6. Enough material is not applied to take into account the tolerances of the parts.

There are some well-developed techniques which should help in the application to various types of joints.

3.1   Straight Threads

A through hole is the easiest space to fill if a material not over 7000 cP is used. Material should be placed onto the bolt or stud in the engagement area and the parts run together at relatively slow hand speed (Fig. 4.44). If a nut or bolt runner is used, then it is preferable to place the liquid into the male part to prevent flinging material at high assembly speed. Generally this kind of production setup can justify the use of dry materials MM, NN, SS, or TT, which have no problem with high-speed rotation.

For maintenance assembly where parts may be very dirty, it is best to apply material to both parts. In any case an excess of material should always be evident at each end of the thread. A quick wipe with an absorbent wiper will prevent migration to other parts.

Assembly of a bolt or stud into a blind hole (Fig. 4.45) always means the pumping or venting of air from the hole. This pumping action can blow the male threads clean even though excess material remains evident on the outside. The material will not wick inward against trapped air, and bubbles of air will retard or prevent cure of the small amount of material that does stay in place. The solution is to place the liquid near the bottom of the thread engagement so that air is freely vented for half of the engagement; for the other half it forces the liquid up the threads towards the outside. On threads smaller than 1/4 in. (6 mm) the bottom of the hole should be filled completely with an amount that will show external squeeze-out after assembly. Application must be done with a spout that reaches to the bottom of the hole; otherwise air will be trapped under a large drop in a small hole.

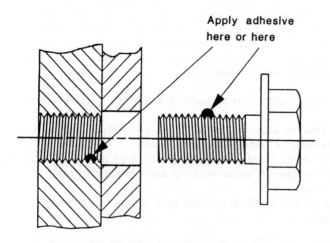

FIGURE 4.44   Straight thread through hole.

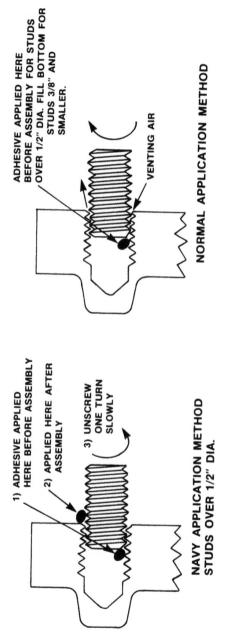

FIGURE 4.45 Straight thread blind hole.

From a design standpoint blind holes should be avoided. With sealed studs and bolts the need for most blind holes is eliminated and the benefits of through holes in manufacturing and assembly can be realized. Through holes are easier to tap and clean when they are manufactured. Further, reassembly of a bolt into a through hole does not require removal of cured material at the bottom of the hole as it might with a blind hole.

The U.S. Navy uses a clever technique on many of the large studs it uses in ship and submarine construction. Where it is necessary to have a blind hole, it is tapped at least a thread deeper than the stud length calls for. Material is applied to the male threads and to the bore. The stud is assembled one turn past the desired position to vent air and some liquid, then it is slowly backed out by hand allowing the excess to be sucked back into the engaged threads. When the assembly must be done overhead then the thicker thixotropic materials are used (e.g., Grade O).

3.2   Cylindrical Assemblies in Blind Holes

Everything said about blind tapped holes applies to blind cylindrical assemblies. Allowance must be made to relieve trapped air. On large-diameter short engagements (Fig. 4.46, left) it is sometimes possible to wick the material in from one side, letting the liquid penetrate inward and gradually spread two ways to the far side, where it joins to provide a seal without trapping air. In critical production applications the material can be applied and the parts placed in a light vacuum. Return to normal pressure forces the liquid into the joint.

3.3   Cylindrical Assemblies in Through Holes

For parts that are 1 in. or less in diameter and have engagements of one diameter or less in length, application to the male part on the bondline suffices to wet both parts if the assembly is made with a twisting motion. It helps to have a leading chamfer on the hole to act as a lead-in for adhesive and a reservoir while wicking completes the fill. It also helps if the female part is wetted with adhesive before assembly. *The surest way to understand the filling process on a production part is to make a model out of clear plastic (methacrylate) and try different techniques.*

When working with thick materials in thin gaps, apply the adhesive to the male part and assemble without twisting or "working" the parts. The opposite may be true with thin materials; twisting helps. One high-production operation that has been running successfully for more than 5 years involves a 0.312 in. (8 mm) shaft in a sleeve that is 10 diameters long. The most successful way to move the material down the bore within the 3-second cycle time was to place the material onto

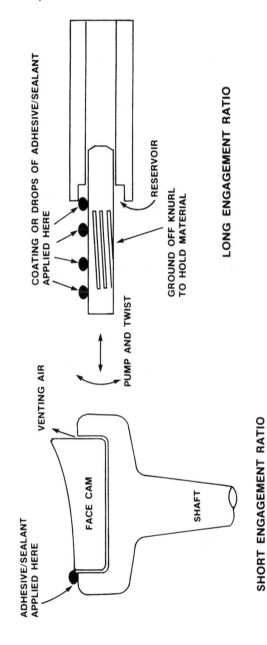

FIGURE 4.46 Cylindrical assembly into a blind hole.

the partially assembled shaft, then, with a programmed machine, to reciprocate and rotate the shaft with a back-and-forth, up-and-down motion for three cycles. Millions of parts have been made without failure. Another method for carrying the material down a lengthy bore is to provide a shallow relief near the center. A shallow relief or series of grooves $0.010 \times 0.125''$ (0.25×3 mm) filled before assembly will often assist in fast penetration. Likewise a light knurling, either straight or helical, will do the same.

3.4 Alignment of Parts

Liquid materials will never align a slip-fitted shaft in a bore. Neither will they cause shaft bending. In general, the alignment of a shaft in a bore with a reasonable sliding fit will be better than a heavy press (which may bend the shaft) or any type of knurl. For the most precise alignment a very light press fit can be maintained. One of the major reasons for using an adhesive fit is to get heavy-duty driving ability without shaft, bore, or bearing distortion. The other reason is, of course, to avoid close tolerance fits.

3.5 Flat Surfaces

Bead Laying

In order to cover flat surfaces and get them assembled without the inclusion of air bubbles, a narrow bead of thick material has proven to be the best. As the bead is compressed by the mating part it is extruded to both sides, pushing the air gap before it. Care should be taken to surround bolt holes or any other interruptions in the surface.

Screening

Anaerobic materials are uniquely appropriate to the screening process, which is adapted from ink and paint printing techniques. They resist drying or curing while exposed in air and remain wet indefinitely on the screen. Care should be taken to control the texture of the print on the parts so that it is not too fine since each nonfilled line potentially represents a trapped air inclusion. Too much trapped air provides potential leak paths and may even prevent cure. Guidelines have been developed by adhesive manufacturers with equipment capability. Their advice and help should be sought.

3.6 Sealing of Tapered Threads

When assembled, tapered pipe threads engage until both thread flanks are firmly in contact. Unlike a straight thread, where the unloaded thread flank leaves a single spiral leak path that must be filled with sealant, a tapered thread has two separate leak paths created by the truncation of the male and female threads. The two leak paths are separated by the firmly engaged flanks; therefore, to seal a tapered thread effectively, sealant must be placed on the male part both on top of the threads and into the root. Then assembly will assure that material is in both the crest and the root, filling the two separated potential leaks. To coat the crest of the thread requires a paste type of adhesive so that it will stay in place. The most reliable is Grade W.

So-called dry seal threads represent an attempt to eliminate the truncation of the threads to avoid using a sealant. Production experience has shown, however, that about 2% will still leak. This condition can be corrected with a very small amount of liquid adhesive such as Grade N.

4. TUMBLING TECHNIQUES

4.1 Introduction

Air stabilized or anaerobic formulas are ideal materials for application to threaded parts by tumbling. The continuous contact with air keeps the material fluid for up to a week as long as parts are not overtumbled or stacked on top of one another. It is important that parts be tumbled only long enough to coat them evenly because metal particles from severe tumbling can destabilize the material in spite of the presence of air. Thousands of parts, quantities depending on size, can be processed in minutes in equipment that is as simple as a plastic bag or a rotating barrel (Figs. 4.47 and 4.48).

4.2 Selection of Sealant Grade

None of the sealants listed in this handbook is as satisfactory for tumbling as the older, slower Mil-S-22473 materials. The materials described in this book should be tumbled in limited amounts and stored for less than 1 hour in single layers.

Viscosity

The best grades for tumbling are relatively thick and slow-curing with a viscosity between 1000 (syrup) to 500,000 cP (paste) (1 to 500 Pa.s). This would include formulas AVV, CVV (Mil Spec 22473D), and W. The safe storage time after tumbling will be related to the published fixture time, the surface being covered, the general cleanliness, and temperature. There are special slow sealants available that can be left unassembled for up to 7 days.

Strength

For sealing, lower strengths are satisfactory (e.g., 400 lb/in.^{2} shear stress). The tumbled material will usually give lower than

FIGURE 4.47   Small cement mixer used to treat large quantities of
pipe plugs.

published shear stress by 20 to 50%, depending on the quantity applied and the amount of residual oil on the parts, which acts as
diluent.   Reference to product performance charts should be made for
shear stress vs. part size and finish.

Coverage

The quantity of sealant required depends on the surface area of
the parts to be treated.   Quantities for common fasteners and plugs
are shown in Table 2.15.   The use of excess sealant should be avoided
as it not only is wasteful but can cause lumps or clusters of parts to
form.   Properly treated parts are merely moist with a thin film of sealant on handling surfaces.   Material tends to concentrate where it is
needed, in threads and under the heads of screws.   Wrench sockets
and heads remain reasonably dry and can be driven by pneumatic
drivers.   Wrenches and driving tools should be cleaned weekly with a
petroleum or chlorinated solvent.

FIGURE 4.48 Central tumbling station prepares and distributes screws for the assembly line.

4.3 Tumbling Sequence

The tumbling operation should include the following steps:

1. Fill the barrel with a predetermined quantity of parts either by weight or volume. In either case, fill only to 1/3 volume to allow for proper tumbling action. (Parts should have been well drained of excess oil.)

2. Turn on the tumbler before adding sealant to prevent premature sticking and spotty application. Parts should slide smoothly over one another without impacting. Quantity of parts and speed of rotation may have to be adjusted to prevent impacting.

3. Add the measured quantity of sealant to the side of the barrel and tumble for a predetermined time.

The time required to achieve a uniform coating varies with the batch size, the part configuration, and viscosity of the sealant. Usually 5 minutes is sufficient. Higher-viscosity sealants may require more tumbling than ones with low viscosity.

By careful observation, driving slots, sockets, or hex heads can be kept free of material by limiting the amount of tumble.

4. Dump parts into clean, nonmetallic pans or polyethylene sacks. Spread in thin layers for longest storage time.

5. Clean the barrel with a clean disposable cloth.

6. Record the tumble quantities and time for processing records.

4.4   Recommendations, Limitations, and Precautions

1. Use of primers or activators is not recommended on tumbled parts since increased cure activity may cause parts to cluster or stick together like popcorn balls. The mating part may be activated to assure fast cure with slow materials.

2. Parts that tend to nest or have flat surfaces that contact, such as hex-headed screws, may require shortened storage time unless they are separated on trays.

3. Plated parts, which normally are slow to activate anaerobic materials, will store longer than unplated parts. Conversely, active metals such as brass or copper, which speed curing, will cure in clusters and should not be tumbled.

4. Fast (30-minute fixture) thread locking materials are generally unacceptable for tumbling. Curing is so fast that parts are sure to stick together and to the container. Automatic liquid application or dry thread lockers should be the choice if speed and strength so dictate.

5. All the handling precautions of the data sheets should be followed, since tumbling is synonymous with high production and continuous exposure. Assemblers should wear finger protection. Most materials are nonallergens unless or until skin trauma occurs. The result of many skin nicks and cuts is a gradual sensitizing of an individual to an allergic reaction similar to that of industrial cutting oils. This situation can be avoided by using rubber gloves or finger tabs. Good hygiene should always be observed. Mechanic's waterless hand soap, followed by water wash, is effective for removal. Material in the eyes should be washed out with copious amounts of water and a physician consulted. Further discussion of safety follows in Sec. 6 at the end of this chapter.

4.5 Equipment

Plating suppliers carry various sizes and shapes of tumbling barrels. Polyethylene barrels are especially desirable since they are completely inert to anaerobics and are easy to clean. Steel barrels can be used with steel parts but a neoprene lining prevents noise and damage of parts.

Measuring of part quantity is done either by volume or with a part-counting scale. The thread locking material can be added by either weight or volume. Some adhesive manufacturers prepackage liquids or pastes in quantities specifically for a one-drum-load application. Liquid handling and metering can also be done by equipment normally used for spot application.

Some high-volume assembly operations are handled by tumbling small screws in plastic bags by hand. This requires only a sturdy bag, a measured amount of adhesive and parts, and a 1-minute agitation of the mixture. Better control with thicker materials and larger parts is obtained with a tumbling barrel (Fig. 4.49).

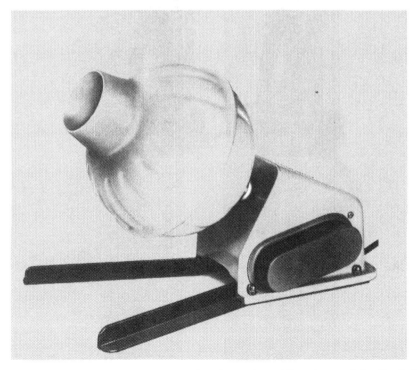

FIGURE 4.49 Bench top tumbling barrels. (Courtesy of A. E. Aubin Co., Manchester, Connecticut.)

FIGURE 4.50   Preapplied Type IV coated bolts.

5. PREAPPLIED MATERIALS

Type VI materials are applied off the line, usually by high-production machinery. The liquid material consists of a water-based or solvent-based slurry, which is either rolled or poured onto the threads of a fastener. The fastener is then heated until the coating is thoroughly dry. The activator or, in the case of two-part epoxies, the second part, is microencapsulated in a fragile shell which is broken during the assembly of the threads. In either case, the fasteners are delivered to the assembly line with a dry band that resembles a soft paint. Assembly into close-fitting threads mixes, liquefies, and activates the chemicals.

Application of these materials is usually done by independent converters who are expert in handling large volumes of fasteners in a precise process. It is recommended that the material manufacturer be contacted for help in selecting a converter.

The cost of screws, bolts, or nuts coated in this manner is usually as low as any other method of liquid application. It yields a total assembled cost of sealing and locking equal to on-line liquids and lower than any mechanical method (Fig. 4.50).

6. SAFETY

6.1 Introduction

Machinery adhesives are based on dimethacrylate esters, their functional component. Mixed into these esters are viscosity modifiers and plasticizers (generally inert polyesters) and fillers to impart the desired physical properties to both liquid and cured products. The formulas also contain a small amount of a proprietary cure system (usually less than 5%), which gives to the product its anaerobic curing properties.

In appearance, the anaerobics are liquids with viscosities varying from water-thin to paste. They are color-coded for identification and visibility purposes and have characteristic odors that are mild and sweet. They have specific gravities of about 1.1 and are largely immiscible with water.

6.2 Overall Toxicology

Anaerobics are considered nontoxic by ingestion having LD_{50} values greater than 10,000 mg/kg (oral, rat). They also have acute dermal toxicity LD_{50} values of greater than 5,000 mg/kg (rat), showing them to be nontoxic by skin contact. (The European Economic Community

Directives regard preparations as not being harmful once the oral and dermal LD$_{50}$ values both exceed 2,000 mg/kg in the rat.) Eye irritancy tests (rabbit) have shown the products to range from nonirritants to mild/moderate eye irritants. The Type I, II, III, V, and VI products are nonvolatile (vapor pressure <0.1 mm Hg/25°C). Type IV products, discussed later, are more volatile because of adhesion promoters (vapor pressure <0.5 mm Hg/25°C). None is subject to spontaneous polymerization.

Machinery adhesives are not primary skin irritants or common sensitizers. While the sensitizing potential of many acrylates is well known, the dimethacrylates used as the principal ingredients of machinery adhesives are, in contrast, not common sensitizing agents. Independent tests have shown negligible evidence of skin irritation or sensitization from the Loctite Type I, II, III, V, and IV versions of these materials. Safety data sheets from the manufacturer should be consulted for any continuous exposure since some formulations include adhesion promoters that change the sensitizing capability of the formulae. A typical Material Safety Data Sheet is shown in Appendix I.

Type IV may typically contain acrylic acid and hydroxypropyl methacrylate. Acrylic acid, like many organic acids, is corrosive in the pure or concentrated form; however, when below 25% concentration, the product is merely an irritant (European Economic Community Directives). The concentration (only in Type IV) never exceeds 10% and usually is 3 to 5%. Hydroxypropyl methacrylate, the other principal adhesive agent used in Type IV, is both an irritant and a sensitizer. Being a relatively small molecule, it can penetrate the skin relatively easily. At over 10% concentration the composition requires the irritant symbol and a warning of the risk of sensitization (EEC Directives).

It should be emphasized, however, that these products, while possessing a slight potential for causing allergic contact dermatitis, are not harmful in comparison to many industrial adhesives and can be handled in complete safety in the industrial environment with minimal precautions. The safety of the materials is supported by accumulated experience over many years of industrial use where continuous or prolonged skin contact frequently occurred without unfavorable effects. Sensitization can occur under some conditions and the following suggestions should be used to avoid them.

6.3 Potential Health Hazards

Sensitization—Predisposition

Experience has shown that sensitization to many otherwise bland chemicals, including machinery adhesives, can occur when the skin is damaged. Such damage may be microlaceration (due, for example, to the repeated handling of sharp threads on screws during an assembly operation) or some form of traumatization that causes repeated bruising of the skin. Other skin damage may be due to the defatting or abrasive action of plant solvents or harsh skin cleaners. Physical or chemical damage to the outer layer of the skin (the stratum corneum) can have a pronounced effect on the potential inward penetration of a variety of molecules, as the depth and quality of the corneum determines the barrier efficiency of the skin.

It may not be generally appreciated that even low humidity in the workplace can dry the skin to the extent that attack from chemicals becomes more likely. Also, damage (physical or chemical) to corneum cells can compromise their osmotic function, even under favorable environmental conditions, resulting in a loss of water, the horny layer's "plasticizer" or softener. This can cause brittleness and cracked skin, allowing the penetration of soaps and detergents which then act as irritants.

Like household soap, there are many relatively innocuous substances (including machinery adhesives) that are safe in normal use but can give rise to allergic reaction when the skin has been predisposed by damage; these include wheat flour, cinnamon, vegetables (especially carrots, parsley, celery, asparagus, lettuce), fruit, meat, leather, cosmetics, antiseptics, Band-Aids, hand creams, and animal hair. (Some of these allergic reactions are described in the following section.) It is easy to appreciate why there are occasionally instances of skin problems while handling anaerobic products in industrial environments, where severe finger bruising or other skin abuse often occurs. Such skin damage may, of course, render the worker more susceptible to contracting an allergy from any one of a number of chemicals he may be handling in the plant (oils, greases, detergents, paints, and so on) or at home. Sometimes, extensive investigations are necessary to establish the true source of allergy in a particular case of dermatitis. Predisposition to allergy by skin damage may also occur more readily in some workers who are more sensitive than others.

Onset and Nature of Allergic Dermatitis

The time to develop allergic contact dermatitis caused by a chemical agent (the allergen or sensitizer) will vary depending on many factors, including the nature of the chemical and the severity of skin damage, and may take from days to years. The condition is often referred to as delayed hypersensitivity. A dry papulovescular or fissured eczema is first noticed at the fingertips with which the chemical (e.g., the anaerobic) has come into contact; this spreads around the fingernails and around the backs of the hands and possibly to other parts of the body through contact by the hands when contaminated with chemical.

Such a rash is uncomfortable rather than incapacitating and normally disappears within a few days to two weeks after eliminating contact with the product.

It is important to understand that such rashes result from the formation in vivo of an allergen-protein conjugate which causes an immunological change within the body; accordingly, such rashes will erupt again as soon as contact is resumed with the allergen, even in trace amounts. This sensitivity can last for years, sometimes for life. In this respect, allergic contact dermatitis can be more serious than a skin irritation reaction, where the inflammation is provoked by direct cell damage where contact with the chemical has occurred without change in body chemistry.

Precautions Against Sensitization (General)

The best precaution is not to become sensitized in the first place. With anaerobics, which are not common sensitizers, this is usually not difficult: it involves preventing the anaerobic from contacting damaged, unhealthy, or otherwise sensitive skin. It must be emphasized again that skin contact, even for prolonged periods, does not necessarily mean that allergic contact dermatitis will ensue; nevertheless, such a risk may exist and adequate precautions should be taken. Use of applicator equipment is recommended, as this eliminates all skin contact. Equipment is especially helpful in situations where the assembly process involves handling sharply threaded, serrated, or rough-cast metal components.

Alternatively, providing an effective barrier between the skin and the chemical may be a useful approach. Since hands are the most common site for occupational dermatitis, disposable gloves may be used. Such gloves must not be subject to tearing or puncturing (thus allowing ingress of liquid and consequently maintaining its contact with the skin) and, naturally, must be impermeable to the liquid. Natural or synthetic rubber, polyethylene, or PVC gloves are effective. Care should be taken that excess perspiration is not induced through overly long use of such gloves, as this can reduce the barrier function of the skin itself and allow more ready penetration of allergens if the hands are subsequently exposed to them.

Barrier creams are sometimes considered as an alternative method of protection. However, these too have attendant risks, such as inducing allergy in sensitive skins or requiring harsher than normal washing procedures to remove, not to mention the probability of contaminating parts as they are handled. Since in many cases it is doubtful that barrier creams actually provide an effective barrier, they cannot be recommended.

While there is no risk of sensitization from fully cured (polymerized) anaerobics, care should be taken in the handling of assembled parts before they are washed or cleaned since uncured material may still be present outside the bondline.

6.4 Other Potential Health Hazards

Air Contamination

The slight odor characteristic of anaerobics should be of minimal health concern. It is due mainly to traces of solvent in the methacrylate monomer left from its manufacturing process. Some local ventilation is advisable if large areas are treated with product and left exposed for long periods. Ventilation should also be considered whenever screen printing of gasketing products is involved, again due to the large surface area exposed.

Spillage

Spillage can be dispersed and cleaned with soap and water or detergent solutions, or with common industrial or dry cleaning solvents. Very small spillage can be mopped up with cloths or tissues. For fine fabrics, soaking spots with vegetable oil before dry cleaning helps remove traces of pigments.

For disposal, run liquid anaerobics to waste, diluting greatly with water. Anaerobics are largely biodegradable before cure.

Fire

Anaerobics are not spontaneously combustible in small shipping quantities and present no fire hazards. The flashpoint of these materials is greater than 100°C. A small volume of toxic gases (oxides of nitrogen) could theoretically be released on incineration. The decomposition products from such combustion should therefore not be inhaled. When dealing with a fire consuming a large quantity of material, breathing apparatus should be worn. The relatively small volumes of anaerobics normally in use or in storage at any time usually means that fire hazards are of little consequence.

Proximity of the products to a fire resulting in excessively high temperatures could induce polymerization of the products within their containers, causing some heat evolution. The exotherm during polymerization is of concern only if the volume of material exceeds 10 liters and the material is very old, or the cooling and aeration has ceased, as might happen in a tank of impregnant. The manufacturer's instructions for avoiding such a contingency should be strictly followed.

6.5 First Aid Information

Skin Contact

After wiping visible material from the skin, lather with a waterless mechanic's hand soap and wash with soap and water. Use of solvents for cleansing skin is unnecessary and not recommended.

Eye Contact

In the event of accidental eye contact, flush the eye thoroughly with gently flowing water for about 15 minutes and seek medical advice. These materials cannot bond skin or eye tissue.

Ingestion

In the unlikely event of a significant quantity being swallowed, give copious quantities of milk or water to drink. Inducing vomiting is not recommended.

Inhalation

At normal temperatures anaerobics are not volatile and no precautions or treatments are called for. Certain individuals may find exposure to the slight odor unpleasant in which case removal to fresh air is all that is required. If anaerobics are exposed to high temperatures (e.g., on contacting a hot surface), a small amount of vapor may evolve which should not be inhaled. In the event that exposure occurs the remedy is to get to fresh air.

APPENDIX

Loctite Corporation
Industrial Group

*MATERIAL SAFETY
DATA SHEET*

I. PRODUCT IDENTIFICATION

Product Name ___ADHESIVE/SEALANT 242___   Part No. __242__

Product Type ___Anaerobic___   Formula No. _____

II. COMPOSITION

Ingredients	% by Wt.	Hazard
Polyglycol Dimethacrylates	60-65	
Polyglycol Oleates	25-30	
Sulfimide	3-5	
Silicon Dioxide	< 2	
Hydroperoxide	< 2	Toxic, Irritant
N,N-Dialkyltoluidine	< 1	Toxic

III. CHEMICAL AND PHYSICAL PROPERTIES

Vapor Pressure __< 5 mm at 80°F__   Specify Gravity _1.1 at 75°F_

Vapor Density _____   Boiling Point _> 300°F_

Solubility in Water _Slight_   pH _____

Appearance _Blue Liquid_   Odor __Mild__

IV. TOXICITY AND HEALTH HAZARD DATA

Toxicity

Oral LD 50 > 5000 mg/kg
Dermal LD 50 > 2000 mg/kg

TLV __Does not apply__

Symptoms of Overexposure
May cause dermatitis on prolonged contact in sensitive individuals.

Emergency Treatment Procedures

Ingestion
Do not induce vomiting. Keep individual calm. Obtain medical attention.

Inhalation

Skin Contact
Flush with water.

Eye Contact
Flush with water.

Personal Protection

Eyes
Safety glasses recommended.

Skin
Rubber or plastic gloves recommended.

Ventilation

V. FLAMMABILITY AND EXPLOSIVE PROPERTIES

Flash Point __> 200°F__   Method __T.C.C.__

Explosive Limits (% by volume in air)   Lower _____%   Upper _____%

Recommended Extinguishing Agents __CO_2, Foam, Dry Chemical__

Hazardous Products Formed by Fire or Thermal Decomposition

Unusal Fire or Explosion Hazards

None

Compressed Gases   Name _____

                 Pressure at Room Temperature _____

VI. REACTIVITY DATA

Stability   ☒ Stable   ☐ Unstable

Hazardous Polymerization   ☐ May Occur   ☒ Will Not Occur

Hazardous Decomposition Products (non-thermal)

Incompatibility

V. SPILL OR LEAK AND DISPOSAL PROCEDURES

Steps to be taken in case of spill or leak
Soak up in an inert absorbent. Store in partly filled, closed container until disposal.

Recommended methods of disposal
Bury or incinerate in accordance with EPA and local regulations.

VIII. STORAGE AND HANDLING PROCEDURES

Storage
Store below 110°F to preserve shelf-life.

Handling
Avoid prolonged skin contact.

IX. SHIPPING REGULATIONS

Type or Class   DOT   Unrestricted

              IATA   Unrestricted

Proper Shipping Name   DOT

                     IATA

Prepared By   Martin Hauser

Title   Vice President - Environmental Health and Safety

Date   September 22, 1982

Chapter 5
Secure Bolting

1. DESIGNING THE BOLTED JOINT

1.1 Getting the Right Clamping Force

When you buy a nut and bolt, with few exceptions you buy just one thing, and that is clamping force. You want to be able to predict what the force is going to be and how long it will stay at that value. In addition, at the end of a period of time, you may wish to remove the clamping force. Nuts and bolts fill this function well, but must be engineered properly to give satisfactory long-term results.

We tighten a screw or bolt by applying a torque to the head. A clockwise torque makes the bolt to nut distance shorter. If a resistance is met (such as in clamping a flange), the bolt will continue to rotate until a balance is obtained between the torque applied to the head and the sum of bolt tension and friction. The distribution of torque among these three factors is shown in Table 5.1.

The equilibrium relationship between clamp load and torque is often expressed mathematically.

$$T = K \, D \, F$$

where T = torque, in.-lb ($N \cdot m$); D = nominal diameter of bolt, in. (m); F = induced force or clamp load, lb (N); K = torque coefficient or friction factor, an empirical constant that takes into account friction and the variable diameter under the head and in threads where friction is acting. (It is not the coefficient of friction, although it is related to it.) Values of K are determined experimentally. See Table 2.6. (The appendix presents a mathematical analysis of torque and friction.)

TABLE 5.1 Torque Absorption in a Tightened Bolt

	Percent of tightening torque	
	Coarse thread	Fine thread
Bolt tension	15	10
Thread friction	39	42
Head friction	46	48
Total tightening torque	100	100
Loosening torque	70	80

The variation in friction, and therefore K, is wide since it is the result of extremely high pressure between surfaces that may be rough, smooth, oxidized, chemically treated, and/or lubricated. Oily steel has a K that varies between 0.11 and 0.17 or ±20%. Friction absorbs 80–90% of the tightening torque (Table 5.1). Therefore, it is prudent to test a particular combination in a torque-testing device (e.g., a Skidmore–Wilhelm Hydraulic Bolt Tension Tester) to determine proper torque values for assuring good control of bolt tension. Technical data for lubricants and other thread treating materials will often have the K values plotted in torque tension curves as in Fig. 5.1.

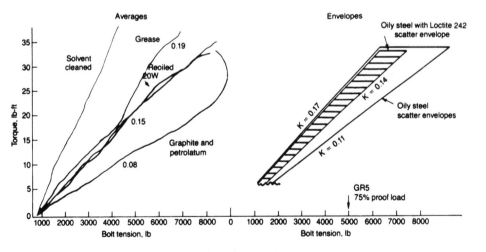

FIGURE 5.1 Torque tension relationships and envelopes, Grade 5 steel.

TABLE 5.2   Typical K Values—Lubricating Thread Lockers on Various Materials[a]

Substrate	Oil	Grades M, N, O
Steel	0.15	0.14
Phosphate	0.13	0.11
Cadmium	0.14	0.13
Stainless 404	0.22	0.17
Zinc	0.18	0.16
Brass	0.16	0.09
Silicon bronze	0.18	0.24
Al. 6262-Ta	0.17	0.29
Dry Degreased Fasteners		
Steel	0.20	0.20
Phosphate	0.24	0.14
Nylon	0.05	0.13
Zinc	0.17	0.15

[a] All specimens were dipped in 5% oil solution and dried before the thread locker was applied (Heat Bath Corp., Lab Oil 72D). Range of values for any lot of fasteners was ±15%, however, different fastener lots can increase the variation to ±20%.

These values were obtained on 3/8 × 16 nuts and bolts where the nut was turned. Both the threads and the nut face were lubricated. An unlubricated thrust surface, either nut or bolt head, can almost double the K value.

1.2   Choosing the Right Bolt

The slope of these straight-line plots, that is, torque divided by tension, is equal to factor K × D and is constant. Since D is also a constant (0.375 in.) then each plot has a computable constant K, where $K = T/FD$. K is the same for all diameters.

  Knowing the friction constant K, the designer can compute the torque tension relationships for other sizes of bolts. To simplify the computation, a nomograph can be drawn as in the right side of Fig. 5.2.

In addition to the graphing of the expression T = KDF, another critical calculation, S = F/A, has been interrelated on the same nomograph where S = bolt stress lb/in.^{2}. With this nomograph a complete fastener design can be approximated for rigid clamping by starting with an allowable stress for the steel grade or material selected. A typical manufacturer's catalog showing allowable stresses for different materials is reprinted in Fig. 5.3. The allowable load on a fastener is based on the proof load or stress. Proof stress is that stress which will not produce permanent elongation. It is just short of the yield stress, which is defined as the stress producing 0.2% permanent elongation. Because of the uncertainty of measuring torque and predicting friction, the design load usually selected is 75 to 80% of the proof load (80% in the example in Fig. 5.2).

Proof loads for individual English and metric bolts and studs are shown in the Appendix in Tables 5.6 and 5.7, respectively. They are calculated from the areas and minimum proof stresses for grades (English) and classes (metric) (Tables 5.8 and 5.9). Manufacturers who wish their fasteners to be classified will certify that their materials meet the grade or class.

Nonrigid joints require further analysis as in Sec. 1.3.

1.3 Gasketed Flange Bolting

The use of a gasket in a joint changes the considerations for bolt selection [2]. The flexibility of the joint increases the possibility that the bolt will experience most of the applied load (Fig. 5.4).

Most of the force for producing minimum sealing pressure in a gasket has to come from the bolts. (A small amount may come from gasket adhesion and gasket swelling from chemical and pressure effects.) It is therefore necessary that the bolts produce the designed stress in the gasket both on initial assembly and throughout its life. There are four reasons why bolts may fail to produce and hold the desired stress in the joint:

1. Bolts that are too large or too small
2. Improper tightening
3. Extreme movement or vibration of the clamped surface
4. Improper material selection with excessive gasket relaxation

Bolt Sizing for Compressible Gaskets

The strength of the bolt must be enough to support the preload, which in turn must be enough to produce the minimum stress that will seal the internal pressure or applied loads. In general, the tension the bolt supports is not increased by the applied load in a solid metal-

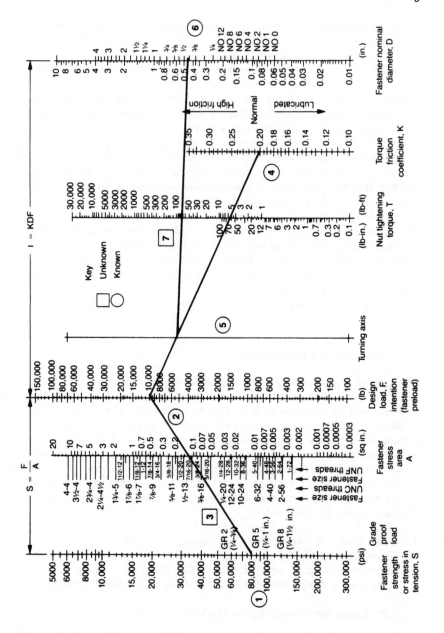

FIGURE 5.2 Threaded fastener monograph. Assume that fastener design load in a bolted joint is 8000 lb. Since friction conditions can vary as much as 20%, assume the 8000 lb represents 80% of clamping force. This means average clamping force is 10,000 lb. Assume the minimum yield strength of the fastener material being considered is 100,000 psi or, using 80% of this value, 80,000 psi. Draw a line between 80,000 psi on S (step 1) and 10,000 lb on F (step 2). The answer, on scale A (step 3), is between 7/16 and 1/2 in. diameter. The next larger size, 1/2 in., should be chosen. To determine the tightening torque required to develop a 10,000-lb clamping load, move to the right side of the nomograph. Assume a normal torque-friction coefficient value of 0.2 (step 4) and draw a line to 10,000 lb on the F scale. Swing around the point of intersection (step 5) on the turning axis to the point on the D scale for a 1/2-in. diameter (step 6). Read required tightening torque (step 7), which is slightly over 80 lb-ft. (© *Machine Design*, from data supplied by the Industrial Fasteners Institute.)

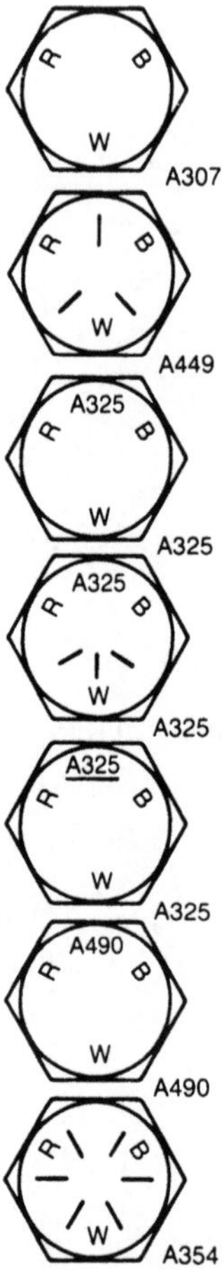

ASTM grade and head marking	Nominal diameter, in.	Proof load, psi	Tensile strength, min. psi
A307	¼-4 ¼-4	Grade A and B Grade B	60,000 min. 100,000 max
A449	¼-1 1⅛-1½ 1¾-3	85,000 74,000 55,000	120,000 105,000 90,000
A325	½-1 1⅛-1½	85,000 74,000	120,000 105,000
A325 Type 1 medium carbon steel			
A325	½-1	85,000	120,000
A325 Type 2 low carbon martensite			
A325	½-1 1⅛-1½	85,000 74,000	120,000 105,000
A325 Type 3 weathering steel			
A490	½-1½	120,000	150,000
A354 Grade BD	¼-4	120,000	150,000

FIGURE 5.3   Allowable bolt stress and head markings.   (Courtesy R.B.W. Co.)

to-metal connection. Certainly the preceding statement is not entirely true, but it is true enough that designers of machine joints can obtain adequate designs based on this assumption [5].

If the general statement of the previous paragraph were true, it would be possible to build an infinitely rigid structure. However, according to Hooke's Law, deflection is proportional to applied force. In most joints the ratio of flange rigidity to fastener rigidity is high enough to discount almost any addition to tension already in the bolt produced by an externally applied load.

However, in a flexible joint with a soft gasket between bolted parts, as in Fig. 5.4, the rigidities of the joint and the bolt are quite different; here, a much greater proportion of the externally applied tension load is added to the bolt preload. The reason for this may become more obvious by studying the following equation:

$$P = P_i + CF_a \qquad (5.1)$$

where P = final load on the bolt, lb (N); P_i = initial preload or clamping load developed through tightening, lb (N); F_a = externally applied load, lb (N); and the constant

$$C = (E_b A_b / L_b) / (E_b A_b / L_b + E_g A_g / T_g) \qquad (5.2)$$

where E_b = modulus of elasticity of the bolt, psi (Pa); E_g = modulus of elasticity of the gasket, psi (Pa); A_b = effective cross-sectional area of the bolt, in.^{2} (m^2); A_g = loaded area of the gasket, in.^{2} (m^2); L_b = effective length of bolt, in. (m); and t_g = gasket thickness, in. (m).

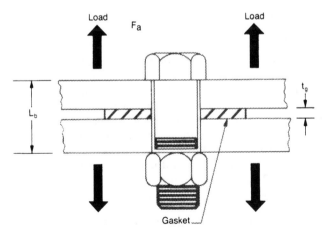

FIGURE 5.4   Gasketed joint.

The value of the constant C falls between 0 and 1. The term $E_g A_g/t_g$ in Eq. 5.2 will be large in comparison to $E_b A_b/L_b$ if the gasket is hard, thin, and large in area. Then the constant C approaches zero. When no gasket is used between members in a rigid joint, C = 0. For very soft gaskets, C approaches 1. It is important to remember that Eq. 5.1 is valid only as long as the gasket remains in contact with joint members. If the bolt stretches to the point where the gasket is no longer in contact, Eq. 5.1 is simply $P = F_a$.

Fatigue Effects

The fatigue strength of a bolted joint must be evaluated from two standpoints: fatigue of the bolt and fatigue of the bolted material. The properly tightened bolt will not fail in fatigue in a rigid joint. Initial bolt tension will stay relatively constant until the external tension load on the joint exceeds the bolt load. Designers do not permit the calculated service load to be greater than the bolt preload. The bolt will experience no appreciable stress variation, and without stress variation, there can be no failure by fatigue, regardless of the number of load cycles on the joint.

This is not the case where considerable flexibility is present. Variable stress in screw or bolt fastenings increases with the flexibility of the connected parts. If flexibility is too great, the variable stress present may be high enough to cause eventual fatigue failure of the fastener regardless of the initial bolt preload.

The greatest single factor that can eliminate cyclic stress variation due to cyclic loading is proper pretensioning or preloading of the fastener. Test results indicate that rigid members bolted together by relatively elastic bolts offer the best method to prevent fatigue failure [5].

1.4   Selection of a Safe Design Load Level

Force Spectrum Analysis

Most designers are well versed in designing for fatigue resistance. A properly clamped joint is designed so that it will not see forces often enough to cause fatigue failures (see Sec. 1.3). A problem becomes apparent when the force spectrum of an application is studied. Fig. 5.5 shows a small segment of a strain gage readout for a joint in an aircraft.

The immediate question is what should be the design load of the joint? We know that a joint clamped tighter than applied loads will not experience cyclic loads on the bolts. Should clamp load be so high that the bolt never sees the 1000-lb peak or can a calculated risk be made to allow a few hundred peaks, which surely could not cause fatigue if below the fatigue limit? The answer is twofold. First, the spectrum should be analyzed for number of occurrences. Fig. 5.6 is the result of such an analysis.

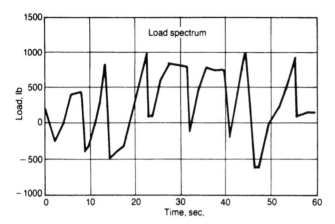

FIGURE 5.5   Aircraft load spectrum.

Second, the other consideration as serious as fatigue must be dealt with. Overloading of a joint will cause self-loosening of the fasteners. Only 50 side slips of a joint will cause 20% clamp load loss [4]. As clamp load is lost, lower and more frequent loads cause slipping, and failure becomes precipitous. Therefore, the designer should consider the bolt securing means at the same time the bolt size is being selected, remembering that securing means affect the clamp load as well as the self-loosening tendencies. If the bolt can be secured against self-loosening, then a few hundred overload cycles will not affect performance. The design load can be a fraction of the peak load, with obvious benefits of cost and weight.

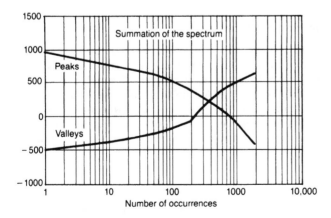

FIGURE 5.6   Summation of the load spectrum.

The proper thread securing means can be selected if the loosening process, as explained in the next section, is well understood.

2. LOOSENING TENDENCIES OF BOLTED JOINTS

2.1 Clamping of Soft Materials

The information obtained from Fig. 5.2 will be satisfactory only for fairly rigid joints. When clamping a "soft" joint, such as a gasketed cover, it is often necessary to determine bolt tightness by the capability of the gasket to support the load [2]. This means that if bolts are chosen that are too large, they will be stressed at a very low value. Since a steel bolt elongates 0.001 in. per inch of length for 30,000 psi stress, a lower stress means less elongation. If a gasket shrinks or creeps under load, the load loss is very rapid when bolt stresses are initially low. For instance, a 2 in. long bolt stressed at 60,000 psi will elongate 0.004 in. If the gasket shrinks 0.001 in., it will lose 25% of its load. If, however, a larger bolt had been chosen that gave the same load at 30,000 psi, then the elongation would be 0.002, and if the gasket shrank 0.001 in., 50% of the load would be lost. The lesson here is to use the longest bolt with the smallest possible diameter that will support the maximum permissible gasket load.

Bolt locking materials, it would seem, do little good if the clamped parts themselves are shrinking away from the bolt. However, whenever there is shrinkage of gaskets, there will also be increased movement of parts in a sliding and longitudinal motion (relative to the bolt). Therefore, a thread locking material or device is beneficial for preventing catastrophic loosening (see Sec. 2.3). If it is a chemical thread filler it also provides thread sealing.

2.2 Brinneling of Bearing Surfaces

Even "hard" flanges and gaskets can collapse under the clamping load if there are burrs under the head or poor finishes on the threads. A relaxing condition can be produced if hard washers are not used under the bearing face of the nut and/or bolt, whichever is to be turned. The primary function of the so-called "lock washer" is to provide a hard bearing surface at very small cost. It has virtually no securing function in spite of its name (see transverse shock and vibration tests).

2.3 Transverse Sliding

All standard bolts and nuts are made with a clearance between them to assure easy assembly. A 3/8 × 16, Class 2A, the most common, will

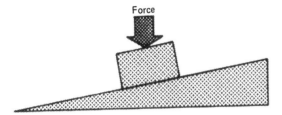

FIGURE 5.7   Nut and bolt schematic of transverse sliding.

have from 0.0013 to 0.0114 in. lateral clearance (M10 × 1.5, Class 6g and 6H will have 0.032 to 0.344 mm clearance). This means that the nut can be moved sideways by this amount. Now consider that the helical thread is nothing more than an inclined plane with the nut sitting on it, held against sliding by friction (Fig. 5.7).

This condition can be simulated by placing a small memo pad or block onto the side of a slippery book (Fig. 5.8). Tip the book until the pad almost slides. Now try to slide the pad sideways with your finger and watch what happens. The pad slides downhill every time you push sideways. You don't have to push it downhill. Its weight moves it by itself. This is exactly what happens to a loaded thread that is made to slide sideways. Each microscopic sideways movement is accompanied by a similar "downhill" motion until the clamp load is completely gone. How fast this happens is shown in Fig. 5.9.

From this experiment and others, we can conclude that if sideways sliding is produced on screw threads, then the threads will unwind all by themselves. The higher the clamping force, the less likely there is to be side movement; but if side movement occurs, the force will unwind the threads regardless of its magnitude.

FIGURE 5.8   Desktop simulation of transverse sliding.

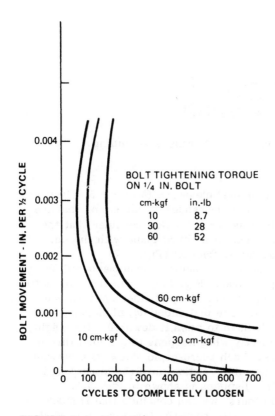

FIGURE 5.9   Loosening rate.

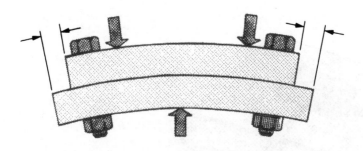

FIGURE 5.10   Bending.

Shear loads and side sliding are common phenomena for bolted assemblies. Side sliding can be caused by bending of the assembly (Fig. 5.10). It can also be caused by differential thermal expansion such as exists on the head of an automobile engine (Fig. 5.11). The differential expansion of a six-cyclinder in-line engine is as much as 0.060 in. total under extreme temperature cycling.

Sliding can be caused by functional loads such as internal pressure, side-loaded pivots, or shock, which is often observed as vibration (Figs. 5.12–5.14).

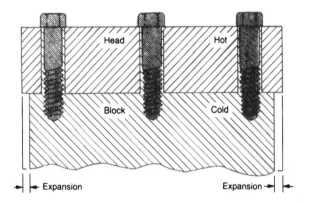

FIGURE 5.11   Thermal stressing.

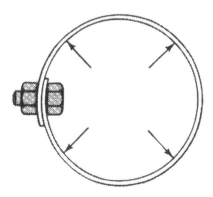

FIGURE 5.12   Internal pressure.

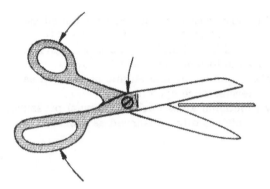

FIGURE 5.13   Side-loaded pivot.

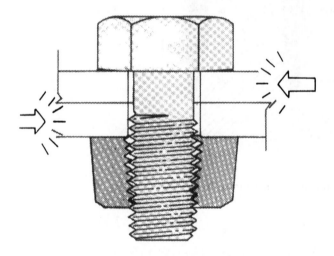

FIGURE 5.14   Shock or impact stressing.

2.4   What About Vibrational Loosening?

Tests done at NASA/Goddard [6] on structures under high vibrational loads of varying frequency substantiated the following conclusions:

1. Vibrational energy had little or no effect on the loosening of a bolt unless side sliding of the threads was simultaneously occurring. Then, and only then, vibration sometimes helped "grease the skids," and loosening proceeded faster than without it.

2. Vibrational energy had a very great effect on the structures being bolted. If the response of the structure caused bending or side sliding, then bolts loosened the same as they did under slow sliding or single impacts.

One structure tested was a simple composite cantilever beam composed of two steel blades bolted together (Fig. 5.15). Slow movement by hand loosened the 1/4 in. bolt after about 100 cycles (as in Figs. 5.10 and 5.11).

Imposition of a vibrational stress on the structure at 1000 Hz and 20 G RMS in an electrohydraulic test machine did not produce loosening in 2 1/2 minutes.

Imposition of a vibrational stress on the structure with mixed frequencies from 200 to 2000 Hz at 10 G produced no loosening, but a 10 G force at 20 to 400 Hz produced bending as in mode 1 (Fig. 5.16) and loosening occurred in 5 to 10 seconds (100 to 200 mode 1 cycles).

2.5 Keeping the Joint Tight—Prevention of Threaded Rotation

Many mechanical devices have been used to prevent unwinding of nuts and bolts. A few such nuts and bolts are shown to indicate the ingenuity which has gone into this effort (Figs. 5.17 and 5.18). Most devices aim to prevent rotary motion and have been successful in a limited way. Worth noting are the nuts with compressible inserts and the nuts with teeth on the bearing surface. The insert is a prevailing-torque type that prevents rotation with a heavy friction drive. By its nature, it resists going on as well as coming off. It does not prevent side motion that limits its effectiveness. The serrated-tooth nut is free spinning and works by digging into the clamped material. It is

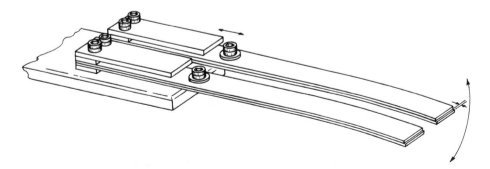

FIGURE 5.15 Bending structure used for testing.

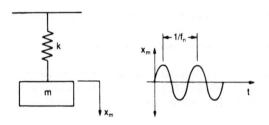

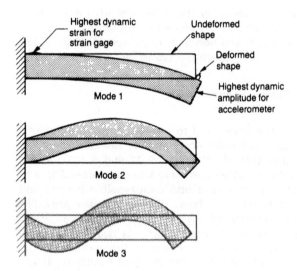

FIGURE 5.16 Vibratory response of a cantilever. (Courtesy of Source Stress Technology, Inc., Rochester, New York.

very effective in preventing rotation. Unfortunately, the digging doesn't stop at the cessation of tightening but continues during use, and on short bolts causes loosening without turning. The digging can also cause failure of the clamped part because it may initiate cracks. It is also possible for either the bolt or nut to turn so both may need some locking features. The cost of mechanical devices exceeds that of chemical thread fillers as shown in Table 5.3.

STRIP INSERT

FUSED NYLON PATCH

THREAD DEFORMATION

DEFORMED THREAD

NYLON PLUG FOR
WEDGING ACTION

NONMETALLIC PLUG
GRIPS BOLT THREADS

FIGURE 5.17   Prevailing-torque fasteners.

SERRATED TOOTH

PREASSEMBLED
WASHER AND SCREW

TOOTHED WASHER

SINGLE THREAD
LOCKNUT GRIP SCREW

FIGURE 5.18   Free-spinning fasteners.

TABLE 5.3   Cost of Securing Compared to Cost of a Plain Steel Bolt

Securing method	Cost (%)
Plain bolt	100
Bolt with liquid thread locker Type I, II, III, or IV	125—137[a]
Bolt plus extra cost of plastic ring lock nut	131—140
Bolt with plastic patch	134—150
Bolt with preapplied Type VI	137—150
Bolt plus extra cost of deformed nut	147—178
Bolt with serrated head	162—202
Bolt with deformed threads	187—206

[a]Includes labor and amortization of automatic equipment in one year.

2.6   Testing with Transverse Shock

The most common and easiest way to test a bolted assembly is by
artificially inducing transverse motion. In the 1960s, G. H. Junkers
designed such a machine, which featured a sinusoidal sliding motion
between two plates clamped with a bolt. He found that as few as 50
cycles would cause a 20% loss in a standard nut and bolt [4]. A similar but simpler machine was designed by a European automobile manufacturer and is the basic machine that the Loctite Corporation used
to evaluate thread locking machinery adhesives. A cross-sectional
picture of this machine is shown in Fig. 5.19. It is called a Transverse Shock and Vibration machine or, more accurately, a trans-shock
machine. It is as capable of rating various locking devices as the
Junkers machine is, even although it is not as versatile. It ends up
with the same relative ratings and more importantly can be correlated
with field results. The test is severe enough to cause failure of most
mechanical locking methods (Fig. 5.20). This doesn't mean that these
methods are not useful to a certain degree; however, on a function
and cost basis they are hard to justify.

From a functional standpoint, the most common mechanical prevailing-torque or thread-jamming types interfere with the proper tightening. Fig. 5.21 shows three individual runs on nuts with a ring
type insert, whereas Fig. 5.20 shows the averages of several specimens. Note in Fig. 5.21 that a 40% variation in torque is needed to
produce the same target clamp load. Also, the better the anti-

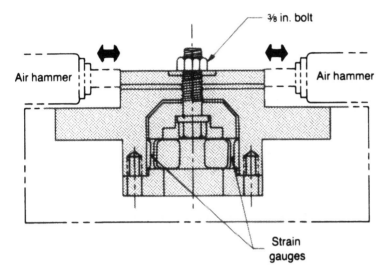

FIGURE 5.19   Transverse shock test machine.

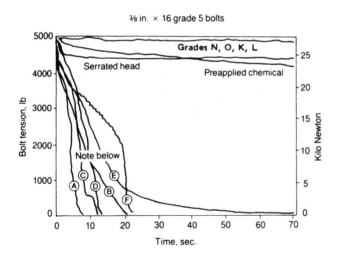

FIGURE 5.20   Results of transverse shock.

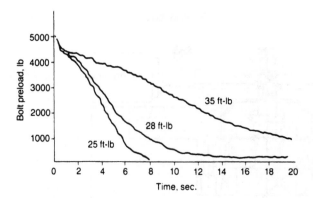

FIGURE 5.21   Effect of prevailing torque on performance.

loosening performance, the less likely that the clamp load will have
been attained. The chief features of mechanical devices are re-
usability, temperature resistance (all metal distorted threads), and
inspectability (Fig. 5.21). The relative costs of some of the common
methods are shown in Table 5.3. A comparison of features is shown
in Table 5.4.

3. PREVENTION OF PREMATURE LOOSENING

The most effective way to prevent unwinding of threads is to prevent
all thread movements in any direction. This is accomplished with
machinery adhesives, which fill and cure in all the open space between the threads, thereby preventing sideways, rotational, tipping,
and dilation thread movement. As one might expect, this system is
most effective as proven by laboratory tests and field experience.
    Selection of the proper material according to its viscosity,
strength and environmental resistance is explained in Chap. 2.

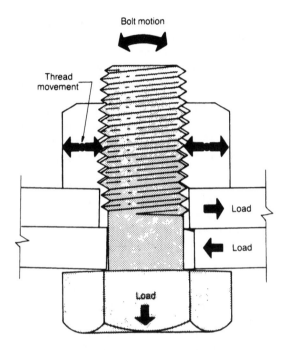

FIGURE 5.22   Filled thread prevents motion.

4.   GENERIC APPLICATIONS FOR THREAD LOCKING ADHESIVES

There are six design situations where the benefits of chemical locking are compelling. By understanding the generic situations, a
designer can avoid failure of structures by specifying the proper
locker before fabrication. Once a generic situation is identified and
treated, other benefits come as part of the application. For instance, a locking need solved with a properly selected thread locker
can make maintenance and disassembly easier in a corrosive situation. The six situations for using chemical lockers are listed below.
   1. To prevent thread movement from bending, thermal, and
functional loads and provide better locking at lower cost than keys,
dutchmen, double nuts, pins, plastic inserts, staking, welding, or
interference fits. For function and cost effectiveness the chemical
lockers are easy to justify. The dry-to-touch materials have overcome application concerns and should be considered whenever the
volume justifies the effort.

2. To protect threads from corrosion and seal leaks in either straight or tapered threads. Materials that cure in place with little shrinkage are ideal for sealing inner spaces, as typified by the helical path through every helical thread. This not only seals the structure but prevents thread corrosion, thus allowing disassembly with predictable torques without fastener damage.

On straight threads, in a shear mode, sealing has been effective at 3500 psi (24 MPa) and on tapered threads, where material is compressed in the helix, pressures of 10,000 psi (69 Mpa) have been surpassed. This is just below the burst strength of iron pipe fittings.

3. To give maximum performance under extreme, life-threatening conditions. Chemicals occupy only inner space so they can be used with other systems to provide redundancy. They work with all types of mechanical lockers with little added cost. They can provide sealing for mechanical devices which seldom seal and often trap corrosive materials.

4. Save space and weight without adversely affecting the metallurgy or strength of the locked parts. On a nut with a crimped collar the weight savings can be the equivalent of the nut weight itself. This results from eliminating the collar and an equivalent length of thread in the bolt. There is no need to compromise the nut or bolt hardness to accommodate yielding elements. Welding, done in desperation, often jeopardizes safety by destroying the fatigue strength of the assembly.

5. To secure threads where the clamp load must be low to accommodate crush or distortion of the joint. When clamping low-modulus material such as gaskets, laminated composites, or plastics, there is a limit to clamp load that is determined not by the bolt but by the clamped material. There will be a tendency to shift under load with the very real possibility of loosening the fastener in only a few hundred cycles. This situation calls for precise development of clamp load and a secure locking means that can operate independently of load. The chemical lockers do both of these well.

6. To give low on-torque and good lubrication in order to generate predictable clamp load. With 85% of the torque being absorbed in friction, a small variation in friction will cause large variation in clamp load. For this reason most formulators of thread lockers have attempted to control friction closely. Grades M, N, and O applied over a light coat of oil give more consistent results than oil alone.

5. SECURING OF STUDS

Studs are widely used to fulfill certain design and assembly functions. Traditionally, they have been difficult to manufacture because of hard-to-control dimensional tolerances requiring critical fitting practices.

TABLE 5.4 Performance of Comparative Locking Methods

	Lubricity	Clamp load scatter	Locking performance[a]	On-torque	Reusability	Thread sealing
Liquid thread locker	Excellent	Low	Excellent	Low	Poor	Yes
Preapplied thread locker	Excellent	Low	Excellent	Low	Fair	Yes
Plastic ring nut	Poor	High	Fair	High	Fair	No
Deformed nut	Poor	High	Poor	High	Fair	No
Plastic patch	Poor	High	Poor	High	Poor	No
Serrated head	Fair	Fair	Excellent	Low	Good	No
Deformed bolt thread	Poor	High	Poor	High	Fair	No

[a]From Transverse Shock Test (Fig. 5.20).

Machinery adhesives are uniquely suited to overcome these manufacturing and performance difficulties.

Studs act as pilots in assembling heavy parts such as turbine casings and cylinder heads. They permit the stacking of parts that form an assembly. The ability of a nut to pivot on the nut end of a stud improves alignment and decreases clamping stress concentration. Clamp load is more easily controlled with a nut than with a tapped hole, especially if the hole is in a hard-to-tap or soft material.

Worn studs are easily and inexpensively repaired, whereas castings with worn or stripped threads require expensive sleeving techniques to restore their function.

5.1   Traditional Problems

In order for a stud system to work effectively, the buried end of the stud must be secured with a greater torque than the nut. The traditional method of doing this has been to assemble with an interference fit, in most cases with selectively measured parts to control the fit. In 1963 a revised standard ASA B1.12 for Class 5 interference fits was established with the purpose of making such a fit more practical, on a statistical control basis, than selected fitting.

Typical problems still persist, stemming largely from the following factors:

1.   The huge variety of analyses and physical and mechanical properties encountered in the forged, die-cast, sand-cast, and rolled materials into which the studs are driven
2.   The inability to closely control sizes of tapped holes in different materials with varying hardness
3.   The variety of materials and heat treatments used for the studs
4.   Variations resulting from rolling, cutting, or grinding external threads
5.   The widely varying effect on friction of chemical coatings, platings, and lubricants

In production use, interference fits, even with the 1963 adjustments, create difficulties involving:

1.   Casting cracking or distortion
2.   Stud shearing
3.   Special taps and gages
4.   Tapping of blind holes to prevent leakage
5.   Controlled lubrication of the tapped hole
6.   Controlled driving speed at high torques

7. High torque power driving tools
8. Counter sinks and starting chamfers
9. Critical depth of engagement to avoid excessive installation torques
10. Loosening due to vibration and shifting of parts
11. Expensive salvage or field repairs after failure

5.2 Machinery Adhesive Method

The adhesive method for studs permits the use of Class 2 or 3 double-end studs with free-running threads at both ends. Studs can be run in by hand with no danger of overstressing the stud or the casting. Grades K, L, or O will cure to a strength that is more closely controlled and usually a higher value than an interference fit will. With all inner space filled, not just the pitch line, resistance to shifting and loosening is superior.

The advantages in using adhesive stud securing are:

1. Free-running Class 2 or 3 fits eliminate shoulder studs and threading from both ends. It permits the use of continuously threaded rod cut to length.
2. Easy to drill and tap through-holes are permissible into sealed pressure spaces.
3. The stud can be installed by hand to a predetermined height.
4. No special taps, gages, or driving tools are required.
5. No special stud-fitting area in the factory is needed.
6. No selective fitting is necessary.
7. No repair process or area is needed.
8. There are no residual stresses in housings.
9. No threads are stripped, cracked, or galled.

It becomes apparent after examining Fig. 5.23 that the chemical fit is more predictable than force fits (see also Table 5.5). The possibility of shearing a stud under 1/2 in. in diameter during assembly is real. With chemical locking the removal torque can be lowered, if desired, by using shorter engagement or a weaker compound.

You must remember that the safe-driving torque for studs is below the recommended tightening torque for bolts. About 45% of the bolt-tightening torque is absorbed under the thrust face of the bolt head or nut. An interference stud resists only with thread friction and is under almost pure torque while being driven. The missing 45% torque resistance of a thrust surface puts the stud near its ultimate strength at what seems like a reasonable torque for a nut or bolt. Safe stud-driving torques will always be below the safe torques for bolts of the same grade.

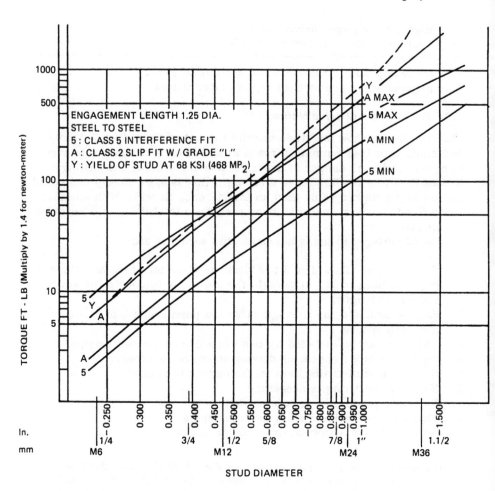

FIGURE 5.23   Torque vs. diameter relationships for studs.

TABLE 5.5   Class 5 Thread
Interferences on Pitch Diameter

Size	Maximum	Minimum
1/4-20	0.0055	0.0003
5/16-18	0.0065	0.0005
3/8-16	0.0070	0.0006
7/16-14	0.0080	0.0008
1/2-13	0.0084	0.0010
9/16-12	0.0092	0.0012
5/8-11	0.0098	0.0014
3/4-10	0.0105	0.0015
7/8-9	0.0116	0.0018
1-8	0.0128	0.0020
1 1/8-7	0.0143	0.0025
1 1/4-7	0.0143	0.0025
1 3/8-6	0.0172	0.0030
1 1/2-6	0.0172	0.0030

With interference fits, safe driving requires closer tolerances and higher strength studs, both steps in the direction of higher cost. Cost studies by a major pump and turbine manufacturer showed that adhesive-assembled studs saved almost two-thirds of the cost of interference studs and their assembly. This saving is typical of converted stud applications.

Other applications are illustrated in Figs. 5.24–5.27.

FIGURE 5.24 Transmission studs in M-60 tanks made by the Chrysler
Corporation are held with Grade K. The Class 5 studs were replaced
with Class 2 studs, cutting assembly costs 75% by eliminating: freezing
studs, heating castings, mixing two-component sealant, selective fitting,
power driving, salvage, and rework of broken studs.

FIGURE 5.25 Anaerobic Grade L is applied to studs used on centri-fugal pipeline compressor. Manufacturer switched from Class 5 to Class 2 threads and cut assembly costs 50%.

FIGURE 5.26 Studs are assembled by hand with Grade L on a large forming press. Previous assembly was done with a 75-lb air-driven impact tool held by two men.

FIGURE 5.27 Grade K is applied to rear main frame studs of TD 25B tractor at International Harvester Plant.

APPENDIX

The Mathematics of Bolt Tightening

A simplified mathematical solution to the torque-tension relationship gives results very close to test results. To start the analysis, one 360° thread segment of a bolt is figuratively unwrapped. The fact that the face is tipped 30° from the plane normal to the bolt axis is ignored. A diagram can then be drawn as in Fig. 5.28. The block represents an element of the nut bearing against the ramp formed by the unwrapped

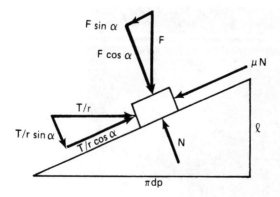

F = Force applied by the bolt

T = Torque applied to the bolt

N = Normal force on friction surface

μ = Coefficient of friction = 0.15

dp = Diametrical pitch

α = Helix angle whose tangent = $\dfrac{\rho}{\pi dp}$

ℓ = Lead of thread

FIGURE 5.28   Free-body diagram of nut and screw thread.

and flattened thread. Since the system is in equilibrium, all forces, with due regard to direction and sign, will balance one another. In other words, all the forces acting parallel to the ramp will sum up to zero and the sum of all forces acting normal to the ramp will equal zero.

$$\Sigma \ F\| \ = 0 \tag{5.3}$$

$$T/r \ \text{Cos} \ \alpha - \mu N - F \ \text{Sin} \ \alpha = 0$$

$$\Sigma \ FN = 0$$

$$N - F \ \text{Cos} \ \alpha - T/r \ \text{Sin} \ \alpha = 0$$

$$N = F \ \text{Cos} \ \alpha + [T/r \ \text{Sin} \ \alpha] \ [\text{a small value that can be dropped}] \tag{5.4}$$

$$T/r \ \text{Cos} \ \alpha - F \ \text{Cos} \ \alpha - F \ \text{Sin} \ \alpha = 0 \qquad \text{(5.4) into (5.3)}$$

or

$$T = r \ (\mu \ F \ Cos \ \alpha/Cos \ \alpha + F \ Sin\alpha/Cos\alpha)$$

$$T = rF \ (\mu + Tan \ \alpha) \tag{5.5}$$

in lb-ft $T = dp/24F \ (\mu + Tan \ \alpha)$

Using this formula examples follow.

3/8 × 16 UNC	3/8 × 24 UNF
F = 5000 lb given	F = 5000 lb given
$\alpha = 3.5°$ $\mu = 0.15$	$\alpha = 2.2°$ $\mu = 0.15$
dp = 0.330 in.	dp = 0.344 in.
F Cos α = N = 4990 lb	F Cos α = N = 4996 lb
N = 750 lb	N = 750 lb
T = 14.5 lb-ft	T = 13.5 lb-ft
Conclusion: A fine thread required less torque for the same force.	

Friction Force Under the Head

Again, with a 5000-lb preload and assuming effective bearing diameter of the nut of 0.400 in., the torque required to overcome the bearing friction is

T = moment arm × force

de = effective diameter of bearing surface (3/8 = 0.40)

$T/r = \mu \ F$

$T = r \ \mu \ F = de \ \mu \ F/24 = 0.4/24 \times 0.15 \times 5000 \ lb = 12.5 \ lb-ft$

Total Torque

UNC	UNF
14.5 54%	13.5
12.5 46%	12.5 48%
27 lb-ft	26 lb-ft

Loosening Torque

In a similar manner, loosening torque can be computed.
Again for F = 5000 lb and μ = 0.15

	UNC 3/8 × 16	UNF 3/8 × 24
Thread loosening torque	6.0 lb-ft	8.0 lb-ft

Conclusion: (1) Fine thread was higher. (2) less than 60% of on-torque

Total loosening (add 12.5 lb-ft for head)	18.5 lb-ft	20.5 lb-ft
Now tightening torque was	27 lb-ft	26 lb-ft

Conclusion: UNC loosening torque is 70% of tightening torque.
UNF loosening torque is 80% of tightening torque

If one assumed that the screw thread was 100% efficient and
there were no friction, then the torque to produce a 5000-lb load
would be:

	4.19 lb-ft	15%	2.7 lb-ft	10%
Or to induce preload		15%		10%
To overcome thread friction		39%		42%
Bearing surface friction		46%		48%

Conclusion: Friction is the key factor using up 85—90% of the total
input.

Tables 5.6—5.9 list proof loads and stresses.

TABLE 5.6 Proof Load for English Bolts

	Coarse threads					Fine threads				
			Minimum proof load (lb)					Minimum proof load (lb)		
Dia.	Thread/in.	Area in.²[a]	Grade 2	Grade 5	Grade 8	Thread/in.	Area in.²	Grade 2	Grade 5	Grade 8
1/4	20	0.0318	1750	2700	3800	28	0.0364	2000	3100	4350
5/16	18	0.0524	2900	4450	6300	24	0.0580	3200	4950	6950
3/8	16	0.0775	4250	6600	9300	24	0.0878	4850	7450	14250
7/16	14	0.1063	5850	9050	12750	20	0.1187	6550	10100	14250
1/2	13	0.1419	7800	12050	17050	20	0.1599	8800	13550	19150
9/16	12	0.182	9450	15450	21850	18	0.203	10550	17250	24350
5/8	11	0.226	11750	19200	40100	18	0.256	13300	21750	30700
3/4	10	0.334	17350	28100	40100	16	0.373	19400	31700	44750
7/8	9	0.462	12900	36050	55450	14	0.509	14250	39700	61100
1	8	0.606	16950	47250	91550	14	0.663	19050	53050	81600
1-1/8	7	0.763	21350	71700	116300	12	0.856	21350	63350	102700
1-1/4	7	0.969	27100	71700	116300	12	1.073	29950	79300	128750
1-3/8	6	1.155	32300	85450	138000	12	1.315	30800	97300	157600
1-1/2	6	1.405	39300	103950	168000	12	1.581	44000	116750	189350

[a]The stress area of threads is computed from the formula $A = 0.7854 (D - 0.9743/n)^2$, where D equals nominal diameter in inches, and n equals threads per inch.

TABLE 5.7   Proof Load (kN)[a] for Metric Bolts (IFI-501)

Nominal thread dia. and pitch	Stress area[b] (mm²)	Class 4.6[c]	Class 4.8	Class 5.8	Class 8.8	Class 9.8	Class 10.9	Class 12.9
M1.6 × 0.35	1.27	—	0.39	0.53	—	0.83	1.14	1.23
M2 × 0.40	2.07	—	0.64	0.87	—	1.35	—	2.01
M2.5 × 0.45	3.39	—	1.05	—	—	2.20	—	3.29
M3 × 0.5	5.03	—	1.56	—	—	3.27	—	4.88
M3.5 × 0.6	6.78	—	2.10	—	—	4.41	—	6.58
M4 × 0.7	8.78	—	2.72	—	—	5.71	—	8.52
M5 × 0.8	14.2	3.20	4.40	5.40	—	9.23	11.8	13.8
M6 × 1	20.1	4.52	6.23	7.64	—	13.1	16.7	19.5
M8 × 1.25	36.6	8.24	11.3	13.9	—	23.8	30.4	35.5
M10 × 1.5	58.0	13.1	18.0	22.0	—	37.7	48.1	56.3
M12 × 1.75	84.3	19.0	26.1	32.0	—	54.8	70.0	81.8

M14 × 2	115	25.9	35.7	43.7	—	74.8	95.4	112
M16 × 2	157	35.3	48.7	59.7	94.2	102	130	152
M20 × 2.5	245	55.1	—	93.1	147	—	203	238
M24 × 3	353	79.4	—	134	212	—	293	342
M30 × 3.5	561	126	—	—	337	—	466	544
M36 × 4	817	184	—	—	490	—	678	850

a Proof load is the load that stresses the fastener less than the elastic limit. No permanent elongation will occur at this load. Convention is to design at 75% of proof load in order not to exceed proof load when torque tension may vary ±30%.

b Stress area = $0.7854 (D - 0.9382P)^2$, where D is nominal thread diameter in mm and P is thread pitch in mm.

c Class numbers are coded to represent the nominal strength. The first figure is the nominal ultimate stress expressed in hundreds of Megapascal. The decimal point and figure after it represent the ratio of elastic limit to the ultimate tensile strength. For example, Class 4.8 has an ultimate tensile stress of 420 MPa and an elastic limit 0.8 times it, or 340 MPa (Table 5.9).

Torque calculation: To calculate the proper tightening torque select a K factor or nut coefficient and calculate the torque from the expression T = KDF. For example, K = 0.20, M14 × 2 fastener Class 9.8, and proof load from table is 74.8 kN. Then: Torque = 0.20 × 0.014 m × 74800N × 0.75 = 157 N·M.

Table 5.8   Proof Stresses for English SAE Grades

Grade	Nominal dia. (in.)	Minimum (lb/in.^{2})
1	Up to 1-1/2	33,000
2	1/4 to 3/4	55,000
	7/8 to 1-1/2	33,000
5	1/4 to 1	85,000
	1-1/8 to 1-1/2	74,000
5.1	#6 to 3/8 sems	85,000
7	Up to 1-1/2	105,000
8	1/4 to 1-1/2	120,000
8.1	Up to 1-1/2 studs	120,000

Table 5.9   Proof Stresses for Metric Classes

Class	Nominal dia. (mm)	Minimum (Megapascal)
4.6	M5 to M36	225
4.8	M1.6 to M16	310
5.8	M5 to M24	380
8.8	M1.6 to M36	600
9.8	M1.6 to M16	650
10.9	M5 to M36	830
12.9	M1.6 to M36	970

REFERENCES

1. *The Heritage of Mechanical Fasteners*, Industrial Fastener Institute, Cleveland, Oh., 1974.

2. R. Andreason, G. S. Haviland, J. Tokarski, *Sealing Technology*, Loctite Corporation, Newington, Ct., 1975, PP16-24.

3. Robert J. Finkelston, "How Much Shake Can Bolted Joints Take?," *Machine Design*, October (1972).

4. J. H. Junker, "New Criteria for Self-Loosening of Fasteners Under Vibration," SAE Paper MO 690055, Society of American Engineers, January, 1969, New York, N.Y.

5. "Fastening and Joining," *Machine Design 14*:83 (1967).

6. J. B. Kerley, "The Use and Misuse of Six Billion Bolts Per Year," Environmental Test and Integration Branch, Engineering Services Division, NASA/Goddard Space Flight Center. Given at 35th meeting of Mechanical Failures Prevention Group, National Bureau of Standards, Gaithersburg, MD, 1982.

7. J. Steele et al., "Coping with Vibratory Stress," *Machine Design*, September 23 (1982).

8. G. S. Haviland, "A Logical Approach to Secure Bolting," Society of Manufacturing Engineers #AD80-329, Detroit, 1980.

9. K. O. Kvernland, *World Metric Standards for Engineering*, Industrial Press Inc., New York, 1978.

Chapter 6
Adhesive Fitting of Cylindrical Parts

1. GENERAL PROBLEM OF INNER SPACE AND HUB STRESS

Whenever two metal surfaces are brought together as an assembly, they must be clamped firmly to produce high friction forces. Even on heavy-force FN-4 fits, the touching is confined to high spots that are limited to 20 to 30% of the total surface available (Fig. 6.1). The rest of the area is a gap of "inner space." Often, to get more contact and higher retaining forces, designers make the hub stresses as high as possible, even to the point of yielding the assembled parts, and finishes and tolerances are reduced to an uneconomical minimum. Instead of increasing the pressure, substantial increases in friction can be provided by filling 100% of the available space with a strong machinery adhesive. A combination of a light locational fit with high disassembly friction can solve many fitting problems (Fig. 6.2).

2. GENERIC DESIGN BENEFITS

Certain generic design situations are benefited by increasing the friction in force fits with the commonly used fitting materials Grades K, L, S, T, and U. The designer can solve problems of assembly, strength and cost by using adhesives to assist the fitting process.

1. Eliminating extra bulk maintained only to achieve high friction force. Thin-walled formed parts are used to reduce weight and cost but are often difficult to press together because of their shape and fragility. In Fig. 6.3, for instance, the parts of an intake manifold for an air compressor are smooth, curved, and difficult to hold. Any high force causes cocking or damaging distortion. Even if assembly is somehow

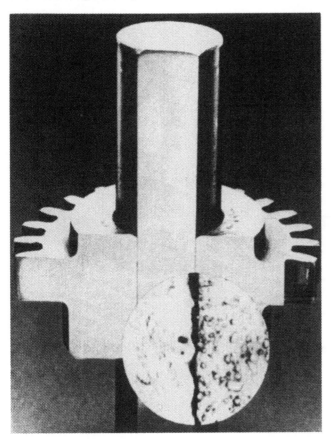

FIGURE 6.1   Cutaway of a gear mounted with a FN-4 press fit.   Insert
1000:1.

accomplished, the tube compresses easily enough so that the assembly
does not stay tight. An adhesive slip fit goes together by hand, is
strong, and provides leakproof security.

Slimming of very strong parts is possible when the effective friction
between parts can be increased as much as four times. For example,
assembling the differential gear in an automobile with a heavy shrink
fit and normal friction produced unsatisfactory strength and excessive
distortion. A lighter shrink fit with a friction-improver similar to Grade
U quadrupled the strength and made regrinding of the gear form unnecessary (Figs. 6.4 and 6.5).

FIGURE 6.2 Machined parts with cylindrical, keyed, and splined fits.

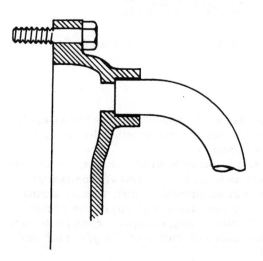

FIGURE 6.3 Close elbow assembled in a manifold with adhesive fit.

2. Augmenting or replacing press fits and reducing cost without changing the design. This is a benefit that should be considered whenever reasonable tolerances cannot produce the friction or holding necessary without damaging distortion. As we will see, this may be true for any press fit with diameters under 1 in. (25 mm). Distortion is caused by heavy residual stress. The press of a perfectly round bearing into a nonsymmetrical housing can destroy the bearing since the housing imposes its lack of symmetry onto the compressed bearing. Or, a slim shaft pushed into a rigid armature will always bend if the fit is tight enough to hold. Restraightening of shafts and repairing galled assemblies is avoided with an adhesive fit.

The tolerances of force fits all lie on the high cost end of the machining-cost curve as illustrated in Fig. 6.6. For instance, USA Standard Force and Shrink Fit tables (USAS B4.1-1967) show that a 1-in. shaft has a tolerance minimum of 0.0009 in. and a maximum of 0.0033 in., which are on the expensive left-hand side of the cost curve. Friction-improvers and retaining compounds allow a slight to major easing of tolerances and costs.

3. Combining materials to get the best qualities of each. Thin brake-cylinder liners are mass-produced from high-wear, corrosion-resistant steel and slipped without distortion into cast housings with an adhesive. Likewise, bronze bushings are mounted in steel structures for medium-speed shaft bearings, and hard drill bushings are reliably assembled in large fixtures of soft, easy-to-cast metal alloys or glass fiber lay-ups (see Sec. 6).

4. Replacing O rings used for static seals. Cured liquid adhesives are superb sealers and can replace static seals under conditions where disassembly is not frequent. Lower-strength materials such as Grade W, X, Y, or Z should be considered so that covers, plugs, or flanges can be easily separated when necessary.

5. Helping compensate for differential thermal expansion. This capability of adhesives is one that can save an application that is borderline with a press fit. Sec. 7 discusses techniques for fitting steel parts into aluminum.

6. Eliminating backlash. Where keys and splines are used for torsional driving without longitudinal sliding, a fit rigidized with an adhesive can improve fatigue life substantially. The micro shifting of keys and splines usually causes fretting and fretting corrosion, which is the stress raiser that starts fatigue cracks. Reversing loads produce shock and vibration, which can cause keyway wallowing and fatigue failure. Impacting can easily double the experienced stress over the normal steady-state load. An example of this was reproduced in the author's laboratory (Fig. 6.7).

(a)

FIGURE 6.4   (a) Differential ring gear assembled with a light shrink
fit and a friction improver similar to Grade U. Interference between
the hardened steel ring gear and carrier is 0.04 to 0.12 mm on a diameter of 140 mm (24 mm wide). Separation force was increased from 5 to
30 tons by using the friction improver. (Courtesy of Renault Cie.)
(b) Automatic application and assembly station for Renault ring gears.

(b)

FIGURE 6.4 (Continued)

(a)

FIGURE 6.5   (a) Teflon coating wheel for applying adhesive, Renault.
(b) Carrier being coated prior to assembly, Renault.

(b)

FIGURE 6.5   (Continued)

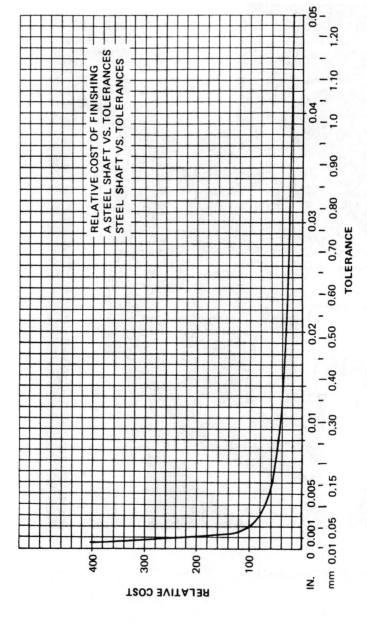

FIGURE 6.6   Costs vs. tolerances for press and adhesive fits.

A snugly fitted Woodruff key in a 1-in. (25-mm) shaft was run
against a key fitted with a machinery adhesive. The 1/4 × 1 3/8 in.
keys were fitted according to SAE J502 standard, which permitted as
much as 0.0008 in. (0.02 mm) clearance but, for this test, were held
to a slight interference fit.

After 12 million reversing cycles on a Sonntag fatigue tester of 150
lb-ft (200 N·m), representing 40% of the shaft torsional shear strength,
the untreated keyway loosened approximately 0.002 in. (0.05 mm) with
serious fretting corrosion and shaft fatigue cracks at the slot. The
adhesively fitted key developed no backlash and showed no signs of
failure.

A practical example of this same effect, shown in Fig. 6.8, is the
spline-mounted gear on a crane/shovel platform, which is used to rotate
a crane on its base. Shaft failures were common until the backlash was
eliminated with an adhesive in the spline. No other changes were needed
to solve the shaft fatigue failure problem.

FIGURE 6.7 Fatigue crack produced in a shaft when coupled with a
tightly fitted key (left) and no failure when assembled with adhesive
(right).

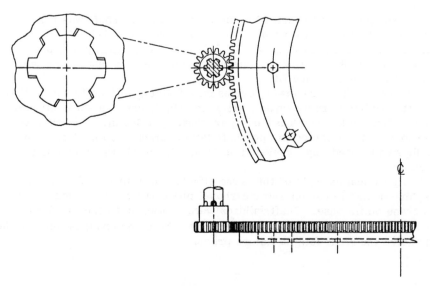

FIGURE 6.8   Movement of splined sprocket caused fatigue failure of
the shaft in a crane turret.

7. Making accurate, rigid assemblies. In the making of sine tables
and jig borer vises, it is necessary to hold accurately ground support
rods in bored supports without distorting the table or support rods.
Sine tables are made to accuracies measured in millionths of an inch.
Press fits and set screw pressure both cause more distortion than allowed. Rods are therefore slipped into place, indicated and adjusted
for accuracy, and then a thin machinery adhesive is applied to cast the
rod into rigid alignment (Fig. 6.9).
    8. Increasing the holding capability of heavy press fits. Under the
extreme pressure and energy of a heavy press fit, most machinery adhesives will cure faster than the fit can be made up. The cured material creates a high friction for the assembly but seems to prevent galling.
If the parts will support the high loads, then this is a good way to improve the load capability of the press. Fatigue limits of the assembly
can be twice those of a plain press. Holding capability can also be
assisted by capillary penetration with Grade R after assembly (sometimes called wicking action). It will achieve surprising penetration on
clean parts and the ultimate strength will achieve five figures on the
heaviest presses.

FIGURE 6.9 Moore sine table and vise. (Courtesy of Moore Special Tool Co.)

3. DESIGN CALCULATIONS

3.1 Press Fits

Tables 6.1 and 6.2 show the results of calculating the holding capability of force fits on shafts of different sizes. The calculation of pressure between hub and shaft is well known from machine design books [1] as derived from thick-walled cylinder formulas (see appendix). To save calculation time, the results of these calculations are plotted in Figs. 6.10–6.12.

Where the hub pressures are known for the interferences and proportions in questions, it is possible to compute the assembly (or disassembly) force or the torque capability from the following formulas:

$$F = \pi \, b \, L \, f \, P \qquad \text{(area} \times \text{friction coefficent} \times \text{pressure)} \qquad (6.1)$$

$$T = \pi \, b^2 L \, f \, P \, /2 \quad \text{(force} \times \text{radius)} \qquad (6.2)$$

TABLE 6.1   Limits of Interference and Part Tolerances for Press and Shrink Fits

Nominal dia. in. (mm)	Tolerance 1/1000 in. (1/1000 mm)		Interference		FN-4 Equivalent (ANSI B4.2) s7-h6 (mm)
	Shaft	Hole	Locational fit FN-2 1/1000 in.	Force fit FN-4 1/1000 in.	
0.1 (2.5)	0.25 (6)	0.4 (10)	0.2-0.85 –	0.3-0.95 –	– (0.008-0.024)
0.2 (5)	0.3 (8)	0.5 (12)	0.2-1.0 –	0.4-1.2 –	– (0.007-0.027)
0.5 (13)	0.4 (11)	0.7 (18)	0.5-1.6 –	0.7-1.8 –	– (0.010-0.039)
1.0 (25)	0.5 (13)	0.8 (21)	0.6-1.9 –	1.0-2.3 –	– (0.014-0.048
4.0 (100)	0.9 (22)	1.4 (35)	1.6-3.9 –	4.6-6.9 –	– (0.036-0.093)
10 (250)	1.2 (29)	2.0 (46)	4.0-7.2 –	10.0-13.2 –	– (0.094-0.169)

where F = force lb. (Newton); T = Torque in.-lb. (N·m); L = length of contact in. (m); b = shaft diameter in. (m); f = coefficient of friction; P = radial hub pressure at the interface of hub and shaft lb/in.^{2} (Pa) (from Fig. 6.10 or 6.11).

In Table 6.2 note that the shafts under one in. (25 mm) in diameter have hub pressures that are excessive (and off the graphs 6.10 and 6.11) at reasonable tolerances. Considerable micro and macro yielding takes place so we calculated the force fit by assuming a maximum hub pressure of 14,500 psi. Practice shows that press fits in the smaller hubs do indeed cause problems of shearing, galling, and shaft bending especially if the hubs are relatively thick and stiff (small b/c). Thin hubs (large b/c) are expanded beyond their yield. Under the conditions producing 14,500 psi hub pressure (e = 0.0017 b/c = 0.65), the tangential tensile stress in the hub is 22,000 psi (Fig. 6.12 extrapolated). Fig. 6.13 shows graphical comparison between hub pressure and tensile stresses in a typical hub.

TABLE 6.2 Limits of Interference and Part Tolerances for Press and Shrink Fits with Assembly Forces

Nominal dia. in. (mm)	Allowance e 1/1000 in./in.	Hub[a] pressure lb/in.^{2}	Ultimate strength	
			FN-4 force[b]	Adhesive slip fit[c]
0.1	6.5	EY @ 14,500	100 lb[d]	160 lb
(2.5)		EY @ 14,500	445 N[d]	700 N
0.2	4.0	EY @ 14,500	410 lb[d]	660
(5)		EY @ 14,500	1.8 KN[d]	3 KN
0.5	2.2	14,500	2600 lb	4100 lb
(13)			12 KN	18 KN
1.0	1.3	11,500	8100 lb	16,500 lb
(25)			36 KN	73 KN
4.0	0.6	5100	58,000 lb	158,000 lb
(100)			260 KN	240 KN
10	0.32	2500	177,000 lb	330,000 lb
(250)			790 KN	1500 KN

[a]Hub pressure read from Fig. 6.10 for allowance e of FN-4 fits. EY = exceeds yield and is off the graph; e was extrapolated for 14,500 psi (approximate hub pressure at tensile yield).

[b]Force required to assemble = π dlfp where d = shaft diameter; l = length of fit assumed to be 1.5d; f = coefficient friction and assumed to be 0.15; p = unit pressure from Fig. 6.10 at d/D = 0.65.

[c]3500 psi up to 1-in. shaft, 2100 psi for 4-in., and 750 psi for 10-in. shaft with 75 in.^{2} and 470 in.^{2} bond areas, respectively.

[d]at yield.

The most variable part of this calculation is the estimate of the coefficient of friction, which from engineering handbooks may vary from 0.10 to 0.33 on steel into cast iron (a 100% variation based on the average).

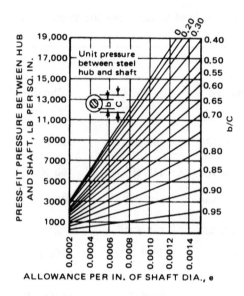

FIGURE 6.10   Press fit pressures between steel hub and shaft. (From Baumeister, Avalione, and Baumeister, *Mark's Standard Handbook for Mechanical Engineers*, copyright 1978.   Courtesy of McGraw-Hill, New York.)

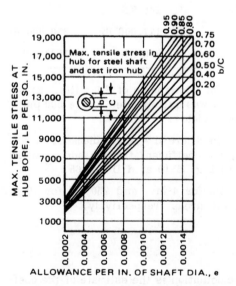

FIGURE 6.11   Press fit pressure between cast iron hub and steel shaft. (From Baumeister, Avalione, and Baumeister, *Mark's Standard Handbook for Mechanical Engineers*, copryight 1978.   Courtesy of McGraw-Hill, New York.)

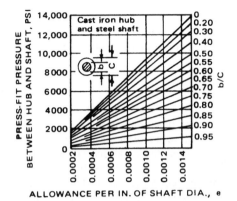

FIGURE 6.12 Tensile stress in cast iron hub versus interference allowance. (From Baumeister, Avalione, and Baumeister, *Mark's Standard Handbook for Mechanical Engineers*, copyright 1978. Courtesy of McGraw-Hill, New York.)

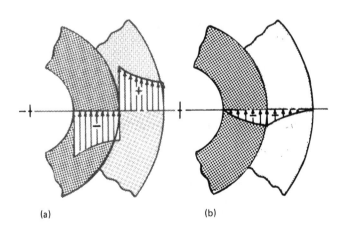

(a)   (b)

FIGURE 6.13 (a) Distribution of tangential stresses in shrink fitted members. (b) Distribution of radial stresses in shrink fitted members. (From J. E. Shigley, *Mechanical Engineering Design*, p. 67, copyright 1977. Courtesy of McGraw-Hill, New York.)

3.2   Adhesive Fits

For adhesive fits, the ultimate strength or disassembly force has been calculated in the last column of Table 6.2 for a modest strength machinery adhesive by the following formula:

$$F = s \times A \text{ (shear stress times area)} \tag{6.3}$$

Over 1 in. diameter the rated shear stress has been reduced 40 and 80% on 4- and 10-in. shafts, respectively, to accommodate the effect of large areas. Studies have shown that the effective average shear strength of bond areas over 10 in.^{2} (6500 mm^2) is inversely related to the area. The reasons for this are purely conjectural, but are probably related to lack of complete fill and stress concentrations more severe than on smaller parts. Under 10 in.^{2} increasing the length of engagement increases not only the area and disassembly force but also the ultimate stress or strength (Fig. 6.14). This means that on small shafts increasing the length 50% can increase the strength 60 to 65%. Rated shear stresses as shown in Chap. 2 were obtained on 0.5 inch shafts with 0.87 engagement ratio and area of 0.69 in.^{2}.

In a similar manner the strength is inversely proportional to gaps over 0.003 in. (0.08 mm). The shear stress used in Eq. 6.3 must be reduced for the size, gap, and fatigue effects as shown in Figs. 6.14–6.16 as appropriate. Steel and cast iron will show the same adhesive properties. Aluminum will be about 1/3 the value of steel (e.g., Fig. 1.19). See Figs. 6.14–6.16.

In summary, then, the acceptable service stress is expressed:

$$s_s = \text{ultimate rated stress} \times \text{service factor/safety factor} \tag{6.4}$$

where the service factor $= f_m \times f_g \times f_s \times f_t \times f_f \times f_c \times f_h$.

Typical Example

Shear rating as obtained on soft steel pin and collar, Grade U = 4500 psi

f_m	material factor, Fig. 1.19, steel on steel	= 1
f_g	gap factor, Fig. 6.15 maximum, diametral clear 0.003 in.	= 1
f_s	size and engagement ratio, Fig. 6.14 L/D = 1; D = 4 in.	= 0.6
f_t	temperature and environmental factor, Fig. 3.3 100°F	= 1.0
f_f	fatigue rating, Fig. 6.16, 10^7 cycles	= 0.2
	safety factor, users practice	= 2
f_c	chemical resistance factor, Table 2.1, air	= 1
f_h	heat aging factor, Fig. 3.16, none	= 1
	Service factor by multiplication	= 0.06
	Safety factor, users practice	= 2

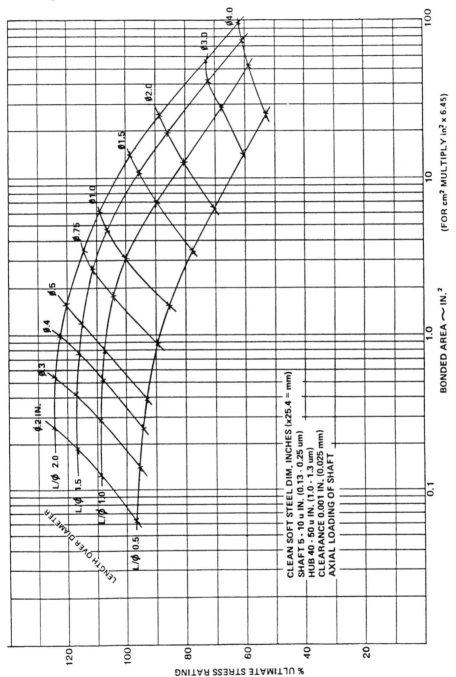

FIGURE 6.14   Size effect on shear strength, strength vs. area.

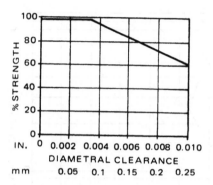

FIGURE 6.15   Strength vs. gap.

Acceptable repetitive shear stress:

   $4500 \times 1 \times 1 \times 0.6 \times 1 \times 0.2 \: / \: 2 = 270 \text{ psi}$

If we had used our rule of thumb of 10% of the ultimate shear rating, the result would have been 450 psi. When possible, the ultimate shear strength should be measured under prescribed conditions so that the interactions of factors for material, size, gap and temperature are exact instead of estimated. Obviously the general factors are conservative

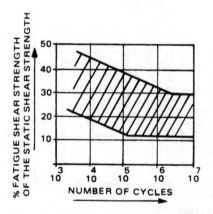

FIGURE 6.16   Fatigue strength vs. cycles.

in their values. Further evidence of their conservative nature is contained in Sec. 4, where we will see that endurance limits on 25-mm shafts are at least 38% of the ultimate strength. This means that, with a safety factor of 2 (as in our 10% assumption), a good design stress is 19% of the ultimate, and, in our example, 855 lb/in.^{2}. Because all data are average, the author would prefer to start with 10% of ultimate as a design stress, which has a built in average safety factor near 4. Functional tests would allow reevaluation.

Surface Finish and Direction of Load

Since the stress ratings are done on pins and collars or nuts and bolts, all with surface finishes less than 64 RMS μin. (1.6 μm), it is possible to increase the stress rating 60% by increasing the roughness to 250 μin. (6.4 μm). The lay of the finish must be perpendicular to the load, or multidirectional. For instance, a pure thrust load is resisted best by circumferential machining marks. A torque load placed on the same parts would show a 40% reduction from the thrust load, coming back close to the original rated load.

The author prefers not to use increased directional roughness of machining marks at the design stage as a means of improving strength. The reasons are the uncertainty of the pureness of the direction of most applications and the existence of two surfaces in every assembly, the less rough of which determines the strength. An exception to this practice is to increase roughness by a multidirectional process such as sand blasting or shot peening. Even then, the less adhesive part is the one that should be improved. The results must be confirmed by test.

3.3 Augmented Press Fits

When one steps back and takes the long view of how to assemble cylindrical parts, it becomes evident that a combination of light or location press and an adhesive for friction improvement can give benefits available to neither alone. These are:

Good location without high residual stress
High ultimate strength exceeding that of either single method
Fatigue strength exceeding that of either single method
Elimination of fretting, corrosion, and leaking
Relaxed manufacturing tolerances and eliminated rework
Simplified assembly tools and fixtures
A minimum disassembly strength, regardless of tolerances, equal to
the adhesive strength

These benefits of adhesives as fitting and friction enhancers have been experienced on fractional horsepower motor shafts as small as 0.1 in. (2.5 mm) with pure adhesive fits, on railroad wheels 5–8 in. in

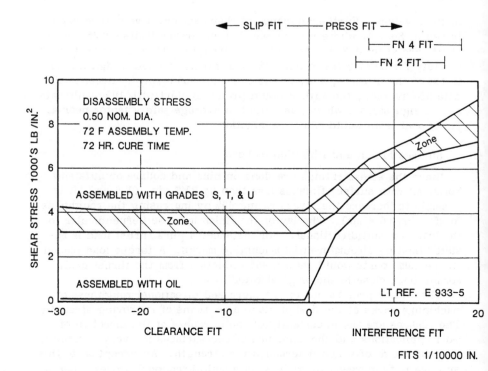

FIGURE 6.17   Shear stress vs. press fit interference for Grades S, T, U, and without adhesive.

diameter (130–200 mm) where adhesive was added to a press, and on a 36 in. in diameter (900 mm) rock crusher tapered shaft 96 in. long (2400 mm) fitted to a chilled cast iron sleeve or roller with a shrink fit.

3.4   Adhesive Shrink Fits

At this time of this writing, experimental work is being done with various types of fits, with and without adhesives.* The results, which are not quite complete, support the following generalizations:

---

*All machinery and mechanical engineering handbooks and mechanical standards of ISO and ANSI list recommended standards for fits of various degrees; they are not repeated in this book.

1. Press fits in steel 3/4 in. (18mm) and under are very difficult to control and usually result in galling or bending of parts. A slip or locational fit (0.000–0.0014-in. interference on 1/2-in. shafts) with an adhesive is the best way to achieve a secure assembly (Fig. 6.17).

2. Press fits 1 in. (25 mm) and over are very beneficially made with an adhesive, although at 3 in. (75 mm) in diameter the ultimate strength can be substantially increased (three-four times) by changing to a shrink fit. At the larger diameters the hub can be expanded enough with heat to give a slip fit at temperature. An adhesive introduced on the cold shaft at this point starts to cure as the parts equilibrate, giving an adhesive layer that is under compressive stress and very strong (Fig. 6.18).

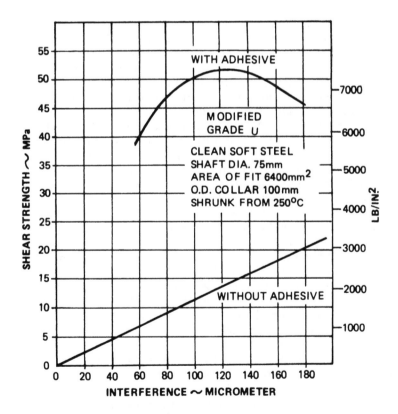

FIGURE 6.18 Shear stress vs. shrink fit interference for modified Grade U and without adhesive.

4. FATIGUE CONSIDERATIONS

It is not enough to stop our calculations at this point and say that the adhesive fit exceeds heavy force fits in force and torque capability. Although this is true for ultimate strength, fatigue strength is usually the design criteria in shaft and hub applications. Since the force fit is maintained by residual stress in the shaft and hub, any applied stresses of torsion or bending must be superimposed on the residual stress [1]. Loss of stress from elevated temperatures or creep can be serious. In small shafts, as we have just seen, the residual stresses are equal to yield, so very careful calculation or tests must be conducted to determine the fatigue life of the assembly and the position of maximum tensile stresses.

Complicating the problems with the press fit may be the stress concentration where the shaft leaves the hub. Although the theoretical value seldom exceeds 2, concentration is dependent on the contact pressure and design of the hub [1]. This concentration of stress often means that failure is incipient at the junction where the shaft enters the hub. This is where slight movement occurs during bending or torsion of the shaft, resulting in displacement of the connecting peaks between the hub and shaft. Oxides are formed by the high pressure on microscopic areas and fretting results with its characteristic brownish-black color. Fretting corrosion can raise the microscopic stress, causing cracking and progressive fatigue failure.

Fatigue calculation of an adhesive fit is less complicated than that of a press fit (Fig. 6.16). It is just a matter of picking the proper reduced stress. Machinery adhesives *on the average* (dangerous, those averages!) will show fatigue strength of 500 psi when tested on relatively stiff shafts and hubs. There are few or no residual stresses, so the calculated stress under functional loads can be used. The rule of thumb used successfully for many years is 10% of the ultimate strength (as adjusted for service conditions; see example in Sec. 3.2), which allows for a 2:1 safety factor and some stress concentration at the shaft hub entry line. Obviously, this represents a considerable downgrading for a material which, at best, is only 10% as strong in tension as the joined materials. However, as we saw on shafts of 1 in. and under, press fits have serious problems of exceeding yield by bending, galling, and expanding to achieve less ultimate grip than the adhesive. Where allowed, extra length can be used to improve an adhesive fit where it is not practical with a press fit because of shaft bending.

Field and laboratory experience indicate that the combination of light press plus an adhesive exceeds the fatigue of either a pure press or pure adhesive. Since all fatigue failures occur in the tensile mode, the compression of the joint in a *modest* way can decrease the possibility that the joint will "see" tensile stresses large enough to be harmful.

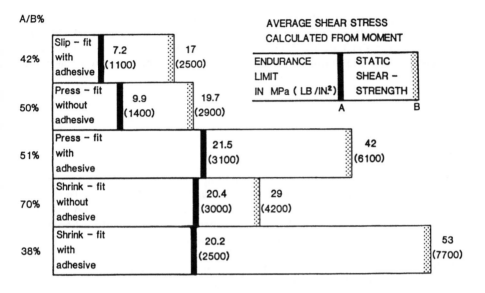

FIGURE 6.19 Endurance limit and ultimate strength for press and shrink fits with and without adhesive.

The laboratory results of a fatigue test done by Steyr-Daimler-Puch in Germany showed that our rules of thumb are most conservative. Fig. 6.19 indicates that fatigue or endurance limits are doubled by adding an adhesive to a press fit but, on shrink fits 25 mm in diameter the endurance is the same with or without adhesive. This would indicate that the ultimate strength cannot be used indiscriminately for the basis of fatigue. In this example fatigue limits of adhesive fits varied from 38 to 50% of the ultimate. It is desirable, as always, to test actual assemblies to determine the effect of residual stress from the fitting process (Fig. 6.19).

5. COMPRESSIVE STRENGTH

The compressive strength of machinery adhesives in thin films far exceeds what one would expect from a tensile strength of 3000 to 6000 lb/in.^{2} (21–41 MPa). In films of 0.003 in. (0.08 mm) loads of 45,000 lb/in.^{2} (310 MPa) can be supported. At this load, soft steel plates start

244 *Fitting of Cylindrical Parts*

to yield and the film will indent the plate permanently. Assemblies that use the materials in compression turn out to be very successful (like a rock crusher shaft in a chilled iron roller). A light press fit plus an adhesive keep the shaft and adhesive in compression in spite of applied loads.

6. BUSHING MOUNTING

6.1 Advantages

Oil-impregnated, sintered bronze bushings can be installed with a slip fit and a machinery adhesive more effectively than they can with a press fit. This procedure exhibits all of the advantages that one can imagine over trying to press fit a slippery, compressible sleeve into a housing that has a different coefficient of expansion:

Holding force is 1.5 to 2.5 times the force of a press fit.
"Close-in" of the bushing bore is eliminated allowing more precise shaft clearance. This means that bushings can be ordered with the inside diameter (ID) to size without the need to perform the extra sizing operation after the parts have been assembled (Fig. 6.20).
The relaxation of housing machining tolerances is possible with resulting cost reduction (Fig. 6.20).
Bushings can be slipped into place without the need for an arbor press.
Improved alignment may be obtained when bushings have a slight clearance fit in the housings and can take their alignment from the shaft. In small motors and sewing machines, this has resulted in freer running and quieter bearings (Fig. 6.21).

6.2 Dimensions and Tolerances

Oil-impregnated bushings are available in a broad range of diameters. Because they are formed in a die, it is possible to hold dimensions closely. See Table 6.3.

This kind of dimensioning is desirable for controlling critical clearances with the shaft. Fig. 6.22 indicates recommended clearances for various applications. For a 1-in. shaft, a maximum diametrical clearance of 0.0015 in. is specified. For smaller, more precise fitting, as with subfractional horsepower motors having 1/4-in. shaft diameters, a diametrical clearance of 0.00075 in. is indicated.

Such clearances require precision grinding of shafts and very accurate control of the ID of the bushing (Figs. 6.20–6.22).

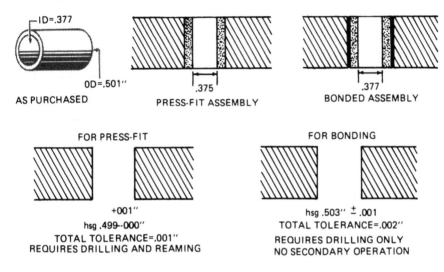

FIGURE 6.20   Typical fitting tolerances for press and adhesion fits.

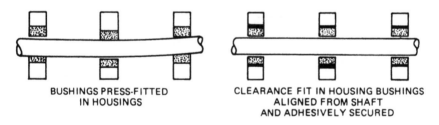

FIGURE 6.21   Alignment of multiple bushings from the shaft.

TABLE 6.3   Typical Bushing Sizes and Tolerances (in.)

Inside or outside dia.	Total clearance	Direction of tolerance	
Below 3/4	0.001	+0.000	−0.001
3/4 to 1 1/2	0.001	+0.000	−0.001
1 1/2 to 2 1/2	0.0015	+0.000	−0.0015
2 1/2 to 3	0.002	+0.000	−0.002
3 to 4	0.003	+0.000	−0.003
4 to 5	0.004	+0.000	−0.004
5 to 6	0.005	+0.000	−0.005

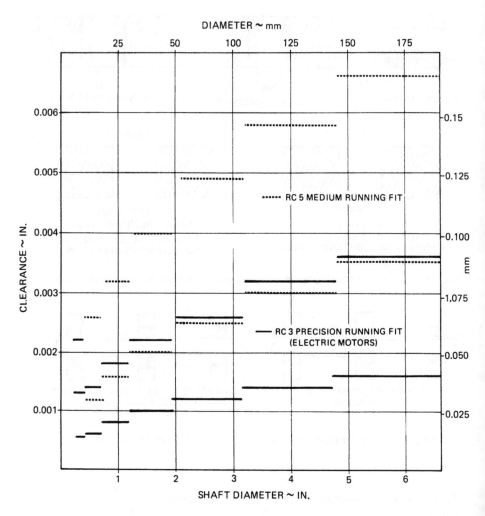

FIGURE 6.22 Recommended shaft clearances vs. shaft diameter.

6.3   Interference Fitting Practice

Interference Allowances

To obtain a press fit, the interference allowance between bushing outside diameter (OD) and housing or hub ID usually falls in the range of 0.0005 to 0.002 in. per in. (or mm per mm) of shaft diameter. For small bushing sizes, it may approach 0.004 in. per in. in order to achieve reasonable tolerances. One manufacturer specifies a press fit of 0.002–0.004 in. (0.05–0.1 mm) on a diameter generally for all sizes.

An interference fit must produce sufficient pressure to hold the bushing in position under all operating conditions. Everything done to make a bushing perform satisfactorily as a low-friction bearing makes it difficult to hold with hub pressure and friction. It must be oily, slippery, and nongalling. Stress produced by the interference fit should not exceed the yield point of the bushing or housing material but it often does, as we saw in Sec. 3, Table 6.2. Excessive stresses are unavoidable if the press fit alone is to be relied on to hold the bushing from moving under functional loads.

Press-Fitting Reduces Size of Bushing Bore

The bore of the bushing closes in approximately in proportion to the interference of the OD with the mating hub when press fitting unless a special installation mandrel is employed.

Sleeve bushings with a 0.501-in. OD when fitted into a 0.498-in. hole close in an average of 0.0024 in. or 80% of the interference on the OD (Fig. 6.23).

If the bore is not resized, the reduction of bore plus assimilation of the housing tolerance results in a shaft-bearing fit that averages 80% looser than the original bushing size allows.

Maintaining the Bushing ID

Resizing of the bore used to be a common practice for restoring the running fit between the bore and shaft. Machining requires accurate positioning of the assembly and extra material on the bore to assure cleanup. Sizing often alters the porosity of the bushing by smearing the pores closed and usually increases the surface roughness. This limits the flow of oil and generally degrades performance of the bearing. The bushing ID is unchanges with an adhesive fit.

Flanged Bushings

When press fitting flanged bushings, there is a greater resistance to close-in at the more massive flange end, with a resulting tapered

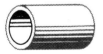

FIGURE 6.23   Sleeve bushing.

condition. Tests have shown that flanged bushings maintain the free
bore size at the flange end, but collapse at the other end with a loss
in parallelism of 0.001 in. (0.025 mm) (Fig. 6.24).

As shown in Fig. 6.22 for a shaft 0.3 in. in diameter, clearance with
bushing should be 0.001 in. Thus, loss in parallelism of 0.001 in. produces either line-to-line fit at the unflanged end or double the recommended clearance at the flanged end.

6.4 Adhesive Mounting Method

By employing an adhesive to bond the bushing in place, it is possible
to open up the hole tolerance in the hub and slip fit the bushing. In
this way, the original bushing ID is maintained. There is no close-in
to make it difficult to maintain specified clearances with shafts. If one
changes a design from press fit to adhesive fit, the shaft diameter may
have to be increased slightly or the bushings bought to finished size
in order to gain the improved shaft-bearing fit.

6.5 Holding Capability

The use of adhesive for retaining, in addition to better dimensional control, also provides superior holding capability. Press-fitted bushings
of 1/2 in. with 0.003-in. interference develop approximately 440 lb average pushout, whereas those treated with Grade S develop 1200 lb.

The presence of oil on the bushing exuding from the capillaries does
not destroy the adhesive's effectiveness. If bushings are heavily flooded
with oil, the excess should be wiped off. The porous texture mechanically augments strength, compensating for light oil films. Cure on
copper-containing bearings is very rapid; however, if immediate testing
is to take place, the hub can be primed with an activator. Some assemblers use activated solvent to quickly wash excess oil off the bushing
just before assembly. In most cases this step is not necessary with
the adhesives listed in this handbook, which are oil-tolerant and fast-
curing.

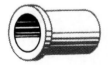

FIGURE 6.24 Flange bushing.

6.6 Heat Transfer

As previously explained, press fitted bushings contact only 20 to 30%
metal-to-metal with their housings. The remaining section of the cir-
cumference is separated from the bore by an air gap, which is an extremely poor conductor of heat. Dead air space is an excellent insulator, with a thermal conductivity coefficient k of 0.015 BTU/hr-ft-°F.
If oil fills the space the conductivity is 0.10.

By contrast, with an adhesive the bushing is surrounded by a hard
plastic film completely filling the area. The conductivity of the plastic
is 0.08 BTU/hr-ft-°F (0.144 W/m/°C) or about five times as great as
that of air and similar to that of oil, although not as high as that of
metal.

Tests of assembled parts have indicated that the heat of friction at
the bushing was readily dissipated into the housing through the adhesive bond. A slip fit with adhesive transferred heat 93% as efficiently
as a press fit, and a light press plus an adhesive was 100% as effective.

6.7 Bushing Lubrication

Bushing lubricants are generally compatible with machinery adhesives
and do not impede their cure or effectiveness. It is not necessary to
wipe off the bushing OD to ensure satisfactory bonding. However, a
few bushing samples should be tested to develop the best technique
and establish that the lubricant will not disrupt processing.

If the bushing oil supply is to be replenished from grooves in the
casting or from supply passages, their characteristic size and air content usually prevents blocking by cured anaerobic adhesive. The oil
seems to find its way into the pores.

6.8 Dry Sintered Bushings

While powdered metal bushings are usually furnished impregnated with
lubricant, occasionally dry bushings are encountered. The porosity of
these parts takes up from 10 to 35% of the total volume. If adhesive is
applied to dry bushings of this type, the pores will act as a sponge to
soak up the liquid adhesive. The effect will be to draw the adhesive
out of the joint, destroying its ability to form a strong bond. Adhesively fitted bushings should be either impregnated with oil before mounting
or secured with nothing thinner than Grade T or Z.

Impregnation of dry bushings can be carried out by heating them,
while they are submerged in oil, to 250°F (120°C) and allowing them to
cool to room temperature while still submerged. Heating expands the
air out of the pores and cooling pulls the oil in.

6.9 Mounting Other Bushing Types

Machinery adhesives are also effective in fitting most hardened plain
or flange bushings as well as nonmetallic Teflon or nylon-filled bushings.
Plastic parts require Primer N treatment to ensure cure.

   Adhesive methods are particularly valuable for securing multiple
drill bushings or bearings into broad thin frames or plates, such as
might be found in drill fixtures or printing frames. Often the cumula-
tive effect of several press fits will distort the frame beyond usefulness. The stress-free assembly with adhesive retains all the original
flatness and hole spacing. Location must be maintained by the fit or
external fixtures since the liquid adhesives have no centering ability.
Nor will the adhesive fit push the bushing off center as is the tendency
with such mechanical fastenings as set screws and tapered keys (Figs.
6.25–6.27).

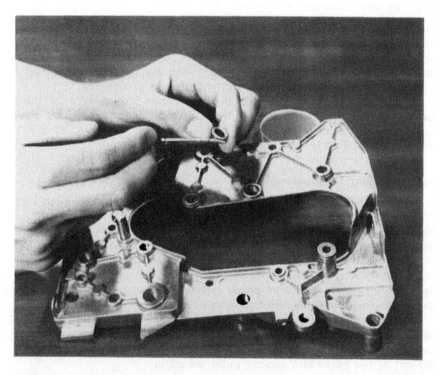

FIGURE 6.25 Typical office machine side frame assembled with multiple
bushings and rods without distortion of the frame.

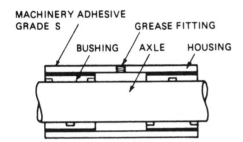

MACHINERY ADHESIVE
GRADE S

GREASE FITTING

BUSHING   AXLE   HOUSING

FIGURE 6.26 The problem of press-fitted bushings being forced out of their housing by grease-gun pressure was eliminated by the use of a machinery adhesive. The method was developed by Parish Steel Division of Dana Corporation for gear-lever axles on heavy-duty trucks. Oil-impregnated bushings could not be press fitted tightly enough to resist the 1700 lb/in.^{2} (12 MPa) pressure of grease guns. The friction improved method required no basic changes in normal assembly procedure.

FIGURE 6.27 Straight sleeve ball bearing bushing die-sets, supplied by Danly Machine Specialties, Inc., use machinery adhesive to avoid close-in, which occurs as a result of a press fit. They are supplied with a wring fit and are retained with the resin. When so installed, the bushing bore does not require honing.

7. MOUNTING OF BALL AND ROLLER BEARINGS

7.1 Traditional Bearing Mounting Practice

Introduction

There exists some controversy about the mounting of bearings using an assist from machinery adhesives. As in most persistent disagreements, the problem is being viewed from more than one direction, somewhat like the blind men trying to describe an elephant by feeling different parts. We shall attempt to look at all sides of the elephant and give the synopsis of 20 years of successful practice.

Before the invention of machinery adhesives, conventional mounting methods used press fits and lock nuts or caps to retain the bearing races. Tolerances on the bearing diameters and fits of the balls into the races were all standardized around press fits and the resulting change of fits as the races were either expanded or shrunk in the process. Reducing tolerances was the standard way to solve loosening and inaccuracy problems. This system has been very successful but, as with any close tolerance assembly, there are cost penalties and failures. Of the eight most commonly encountered failures [2], four are a result of the mounting process:

1. Bearing overload caused by heavy press fits or out-of-round expansion of housings. Out-of-round can be caused by a very slight galling or pickup as a heavy press is consummated. This often remains undetected until the bearing fails prematurely in service.

2. Brinelling caused by improper pressing of the bearing. The press fitting load should not be transmitted through the balls; however, contingencies often make this expedient. A heavy press then destroys the bearing.

3. Misalignment caused by bent shafts or crooked or scored bores. Heavy press fitting practices are not compatible with precision bearings.

4. Cam or inner race fastening methods that cock the bearing. Cam rings and set screws are often used for fastening the inner race. An undersized shaft or overtightening to prevent slip can apply eccentric loads on the bearing.

The other four causes of bearing failure are related not to mounting technique but to environmental effects: contamination, false brinelling caused by vibration when the spindle is stationary, thrust overload, and electric arcing caused by static electricity or faulty wiring.

The Case for Keeping the Press Fit

Bearings are used to locate pulleys, rollers, gears, and cutting tools in a precise position while allowing rotation of parts or spindle around a predetermined axis. They must provide extreme stiffness in

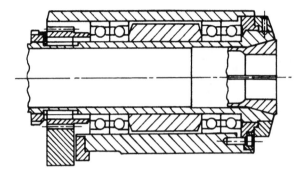

FIGURE 6.28   Typical machine tool spindle.

both axial and radial directions while allowing almost frictionless rotation. In most applications the inner member is a rotating spindle. To
this is attached by a press fit the inner race of the ball or roller bearing
(Fig. 6.28).

There are four reasons given by the bearing manufacturers or users
for maintaining the press fit of the inner race.

1. Concentricity with the spindle is most easily attained by a snug
fit. A heavy press is not necessary but some press is. No centralizing
is achieved with a liquid adhesive. This is a most persuasive argument
which is ignored only during some repair operations, when the cost of
machine down-time makes it worth extra effort to align the bearing on
a worn shaft with shims or fixtures rather than by plating and regrinding to restore the press.

2. The inner ring can rotate under load. The effect of load on a
loose race is a condition shown in Fig. 6.29. Load on the inner ring
leaves space at the bottom of the ring. If this shaft were 1.000 in. in
diameter and the clearance under load 0.001 the two surfaces would act
like gears (as in Fig. 6.29) where the shaft has 1000 teeth and the ring
1001 teeth. As the shaft rotates one revolution, or 1000 teeth spaces,
the ring rotates 1000 positions less one. In other words, the ring
creeps backwards. This action creates friction, fretting corrosion,
wear, and noisy operation. The firm support of a press fit preloads
the mating surfaces and prevents rotation in most cases. However, it
must be remembered from previous discussions of press fits that a
heavy press produces only 30% contact of the touching parts so it is
not difficult to see why loosening can occur under repeated loads. If
this were the only reason given for a press fit then machinery adhesives
would always be used to avoid one of the four common causes of failure.
A light press and a machinery adhesive friction improver will hold at
least twice as well as a very heavy press.

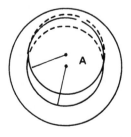 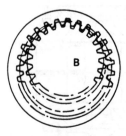

FIGURE 6.29 Effect of load on loosely fitted shaft leaves space below shaft. Rotation under load creates gear effect to cause inner ring to rotate backwards one tooth per revolution with respect to the shaft.

    3. On very heavily loaded inner rings with a premium on space to locate the bearing, such as in a steel rolling mill with tapered roller bearings, the inner race is relatively fragile and must be "rounded up" by the fit. This and drawn cup bearings are the only ones where the bearing roundness and quality do not far surpass the quality of roundness achieved by typical spindle manufacture.
    4. The last reason given for the press fit on the inner ring is the success of years of tradition and the dimensional standards that exist. It is stated that on some ball bearings the fit between the balls and the two races is partially controlled by the expansion of the inner race during the press. A press fit is not a reliable method of reducing the clearance in a bearing since it does so only on an average or mean basis and not throughout the spread of tolerances. The best way to reduce clearances is in the manufacture of the bearing. The Anti-Friction Bearing Manufacturers Association (AFBMA) suggests four different fits.

18 to 24 mm diameter	Maximum clearance 0.0000 in.	AFBMA No.
Snug	4	2
Standard	8	0
Loose	11	3
Extra loose	14	4

The Case for Slip Fits

According to some manufacturers, the outer race, under heavy side loading, should intentionally be left loose ("thumb fit") or very lightly pressed in order to allow rotation similar to that shown in Fig. 6.29. The reason is that if this outer ring is kept stationary, all the repetitive ball-rolling load is taken by one side of the ring and fatigue life will be limited. In steel mills where roller bearings are used, outer cups are sometimes press fitted for rigidity but the races are marked so that they can be reassembled in a different position each time the rolls are disassembled for dressing, which is quite often. Obviously, these procedures of under-load rotation or assembly rotation are practical only under very special circumstances.

Although the above reasoning may be sensible for large roller bearings, most ball bearings are in the range of 18 to 50 mm. Here the most sensitive fatigue member is the inner race because of its convex curvature in contact with the balls. In addition, the control of outer ring rotation by relative looseness is not reliable and can lead to severe fretting of the shaft or bearing (Fig. 6.30). Most users will press the outer ring to one degree or another, with the following important exception.

FIGURE 6.30 Fretting failure, the result of mounting bearings loosely.

Longitudinal Slip

Most spindles incorporate at least two bearings separated by enough space to give bending stiffness. The front bearing is as rigidly fixed as is possible both radially and axially. The rear bearing is usually floated with a light slip fit so that the two bearings do not thrust load each other as they are assembled or when temperature changes occur.

Example: A shaft 24 in. long operating at 100°F higher temperature than that of the housing will expand 0.000006 in./in./°F. At a strain of 0.0006 in./in. and a modulus of 30×10^6 lb/in.^{2}, this translates into a theoretical stress of 18000 lbs/in.^{2} (s = eE). With an area of 0.785 in.^{2} the force is 14000 lb. A typical 1-in. radial bearing is rated at 900 lb thrust.

A temperature differential is quite normal for operating spindles and is especially severe in electric motors where electrical loads end up as heat. One bearing is always floated and the most practical place is the largest diameter of the rear bearing.

Floating Inner Rings for Stationary
Spindles or Axles

Where wheels revolve around stationary axles, such as in automobile nondriving wheels or in idler pulleys, the inner ring is floated and the rotating outer ring is assembled with a press. The generalization can be made that the rotating ring is always secured with a press and the stationary ring may or may not be, depending on the requirements for rigidity and axial sliding.

Precision and Light Multidirectional Loading

Machine tool spindles are often loaded in many radial and axial directions. This requires that the races be as rigidly and concentrically mounted as possible. Drill presses, jig borers, portable tools, and internal high-speed grinders all rely on the rigid and concentric mounting of the front bearing(s) to achieve repeatable positioning and to perform without chattering. In most instances, the front bearings will be axially preloaded to remove all slack from the balls and races. This is done either with spring loads or with pairs of preloaded bearings that are manufactured as pairs with built-in preload (Fig. 6.31).

In cases like this, both the inner and outer race of the front bearing must be very rigidly mounted but press fits must be light so that the preloaded bearings are not overstressed. Since the light fit is satisfactory for the light loads encountered, adhesives or friction improvers are not used since spindle rebuilders say "they give an impression of sloppy fits" and are used only when press failures occur and emergency

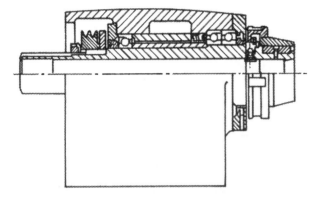

FIGURE 6.31   Precision lathe spindle with spring-induced preload on
the bearings.

repairs are made. Transitional fits are used according to the requirements for concentricity; thus some degree of interference is achieved
on the average.

Shaft and Housing Fits

Realizing the need for different types of interference and clearance
fits of bearings, the International Standards Association (ISA) has prepared standards for them. The internal clearances of the bearings are
designed with specific amounts of interference or looseness depending
on application conditions:

Bearing size and type
Amount of load
Type of load, radial, or thrust
Operating conditions
Possible rotation of load

A graphical representation of the tolerances and allowances for a
j5 drive fit is shown in Fig. 6.32. Note that with a mean fit of 0.0002
in. (0.005 mm) the extremes are from loose 0.0002 to tight 0.0006 in.
(0.005 to 0.013 mm). The extremes can be very troublesome. When
the entire load environment is known and all of the tolerances are carefully controlled the bearing should operate satisfactorily.

Machinery adhesives as friction improvers can enter this scenario
with beneficial results by allowing lighter fits and making loose fits
work, provided concentricity requirements are met.

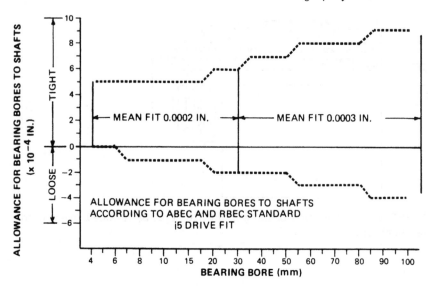

FIGURE 6.32   Allowance for bearing bores to shafts according to ABEC
and RBEC Standard j5 drive fit.

7.2   The Generic and Traditional Places for Machinery
Adhesives for Ball Bearing Mounting

In the 20 years that machinery adhesives have been used for securing of bearing races, the uses can all be categorized as augmenting or
assisting the interference fit.   In this role they prevent movement and
fretting, improve alignment, lower residual stresses, lengthen life, and
solve thermal problems.   Friction improvers traditionally have been used
for:

1. Augmenting press fits in very heavy-duty applications where
heavy presses will yield the housing, bend the shaft or slip under load.
Applications: Bearings for differential gears, truck and automotive
transmissions, heavy-duty pumps, and railroad ballast tampers.

2. Augmenting press fits into low-modulus, high-thermal-expansion
materials such as aluminum and engineering plastics. Applications:
Bearings for automobile transmissions, portable tools, automotive pumps,
and aircraft control rod-ends.

3. Augmenting press fits where there is no room for conventional
retainers such as nuts, snap-rings or caps. Applications: Bearings
in conveyer rollers, idler pulleys, tension arms, and textile spindles.

4. Providing alignment and security where long or fragile shafts
cannot be pressed without disturbing accuracy. Applications:
Bearings for typewriter platens, mail-sorting machinery, tape drives,
and office machinery in general.

5. Providing fast and ostensibly temporary repair of worn shafts and housings where restoration of press fits may cost thousands of dollars in down-time and repair costs. Machinery adhesives return the machine to duty in hours for pennies. Applications: Bearings for paper driers, steel rolling mills, farm equipment, mining equipment, and oil drilling machinery.

Augmenting Heavy-Duty Fits

As was seen with press fits in general, the adhesive assist can increase the holding ability two to six times. It is not necessary to use the strongest material. In order of increasing strength, Grades M, N, S, Z, and U are appropriate for securing a bearing. Grade S is the best to start with because it is thin enough to wick into the surface finish. It can even be applied after assembly, although Grade P will be faster-wicking.

With all of these fast materials there may be some curing during the press. Grade S will repair itself if cure starts during the press. As long as the press is not stopped partway, the partial cure does not change the coefficient of friction and final strength is not affected enough to be ineffective. Grade Z is so thick that it stays where it is put and can be applied to small areas for limiting the total strength. One might call it a chemical set screw.

As with any heavy press fit the design must allow access for removal. A screw or hydraulic jack is the best way to disassemble a bonded bearing. Hammers and heat are destructive.

Augmenting Press Fits in Aluminum and Plastic

Introduction

Demands for lighter weight and greater efficiency in vehicles have led product designers to make greater use of aluminum and reenforced engineering plastics. This section will concentrate on the properties of steel in aluminum fits for simplicity's sake. Everything said for steel and aluminum is germaine to any combination of parts with differing moduli and coefficients of thermal expansion.

The problem is threefold: press fits reach the yield of the low-modulus material before adequate hub pressure is attained; change in temperature varies the hub pressure; and the coefficient of friction between dissimilar materials often is rather low. Aluminum, with all of its strength and lightness, shares these drawbacks when combined with steel.

Historical methods for dealing with this problem included the use of heavy shrink fits, cast-in-place steel sleeves, adhesives, and, most commonly, mechanical retention rings. We will discuss the designing of press and shrink fits in combination with adhesive augmentation. A beneficial synergism results when using the two methods. They can achieve strength levels unattainable with severe shrink fits, at the same time reducing the stress levels in the aluminum (Fig. 6.33).

Performance of Press Fits

Steel into aluminum press fits are shown as calculated in Fig. 6.34 for temperatures ranging from −30 to 350°F (−20 to 177°C). This family of curves was generated for a 1-in. steel shaft pressed into a 2-in. aluminum hub. Interference allowances shown on the curves are for a temperature of 70°F (design equations are in the appendix). Later tests of actual press fits showed about 25% lower results, probably because the coefficient of friction used for the calculation (0.12) was too high and because a few "tenths of a thou" error in measurement can have substantial effect. This disparity between calculated and actual results highlights the problem of predicting press fit performance from pressure and an assumed coefficient of friction. Small changes in either the fit or coefficient are directly reflected in performance.

Limiting hub stresses: Using the information in Fig. 6.34, it would seem possible to select an interference value that would provide any desired level of joint strength needed. Hub stresses, however, severely limit the choice of press fits.

Fig. 6.35 shows the maximum hub stresses that would occur for various temperatures and levels of interferences shown in Fig. 6.34. The

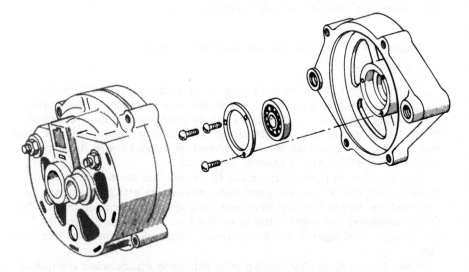

FIGURE 6.33 Steel ball bearing mounted in an aluminum housing with a retaining ring.

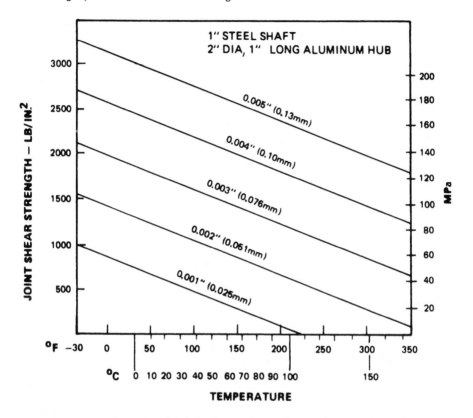

FIGURE 6.34 Calculated joint shear strength vs. temperature for steel/aluminum interference fits.

hub stress value serves as the limit to the amount of interference that can be used. The yield strength of a typical aluminum casting is shown as the limiting stress in Fig. 6.35. This limit is 24,000 lb/in.[2] (165 MPa) tangential shear stress, which translates back to a joint shear strength at a coefficient of friction of 0.12 to 1700 lb/in.[2] (12 MPa) (Eq. 6.9 in appendix). No matter what we do with the press fit we are always limited to a maximum shear strength of 1700 lb/in.[2] or less even with a hub ratio c/b of 0.5, which is very heavy. If we increase the fit beyond this, the hub yields or may split.

We can now replot Fig. 6.34 in a more realistic fashion, as shown in Fig. 6.36.

Any interference over about 0.002 in. (0.051 mm) when carried down to -30°F (-40°C) will yield and, if returned to higher temperatures, goes down the maximum strength line, where at 350°F (180°C) its strength is reduced to about 400 lb/in.[2] (3 MPa). If we had corrected the curves according to the actual values, which were 25% lower, the 350°F fit would be zero.

Hub yielding and low friction should be enough evidence to show why press fits are not practical, and the consideration of tolerances will show that the theoretical fits cannot be controlled. A tolerance of ±0.0005 in. (±0.01 mm) on the shaft and ±0.001 in. (0.03 mm) on the hub with a nominal interference of 0.002 in. (0.05 mm) means that the actual interference would range from 0.001 to 0.003 in. (0.025 to 0.08 mm). Some of these parts will loosen at 150°F (76°C).

Functional loads have not been considered. They, of course, must be added to the fitting stresses, the total of which must remain below the yield.

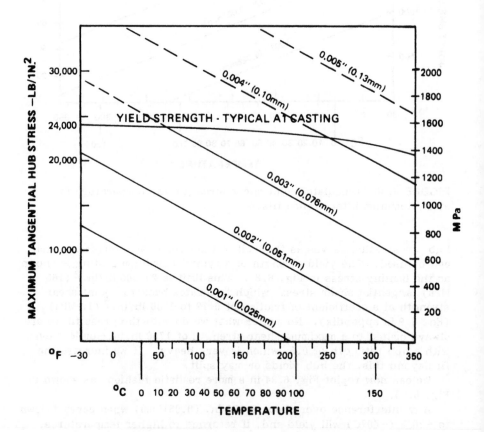

FIGURE 6.35   Calculated hub stress vs. temperature for steel/aluminum interference fits.

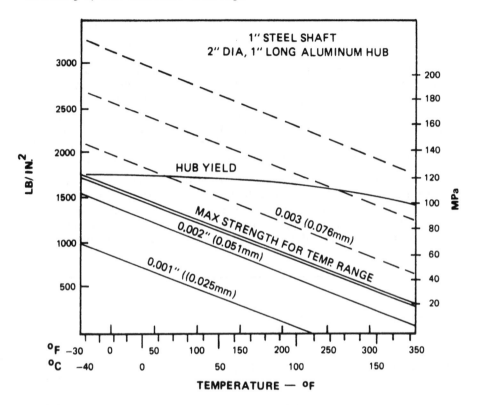

FIGURE 6.36 Joint shear strength vs. temperature for steel/aluminum interference fits.

Use of adhesives to alter the coefficient
of friction

As we have seen, the retention strength is highly dependent on the coefficient of friction, which is notoriously unpredictable. It may change with surface finish, oiliness, temperature, and pressure. For aluminum to steel the frictional coefficient is not only variable but low. Values of 0.10 to 0.15, or 50% variation, are often used for design. The use of adhesives to improve the friction at the interface makes sense because we have 70% of the surface as unused "inner space" until the adhesive is introduced. Filling the inner space with a strong adhesive did improve the strength by a factor of two (Fig. 6.37).

Three adhesive techniques are described below.

1. Adhesive with clearance: This is the conventional bonding method using 0.002 in. (0.051 mm) clearance with machinery adhesive Grade U. Both components were solvent-vapor cleaned prior to assembly.

One-inch shafts in 2-in. diameter by 1-in. long hubs were used
throughout. From prior experience we know that clearances could
have varied between 0.001 in. (0.025 mm) and 0.005 in. (0.13 mm)
without affecting the strength. The shear strength at room temperature (at which it was cured) was 1400 lb/in.^{2} (10 MPa). As the
parts were heated the aluminum grew away from the shaft and the
adhesive. Even though the coefficient of thermal expansion of the adhesive exceeds that of the aluminum, the adhesive film is so thin that
its total expansion is relatively low and the film begins to fail in tension.
Performance was equal to or better than that of the press fit.

    2. Adhesive with interference: Grade U was used to augment a
0.002 in. (0.025 mm) press fit with exactly the same preparation as
above. Strengths for this method were more than double those of the
press fit. The interference fit prevents separation of the surfaces so
the adhesive is always effective. The adhesive bond adds joint strength

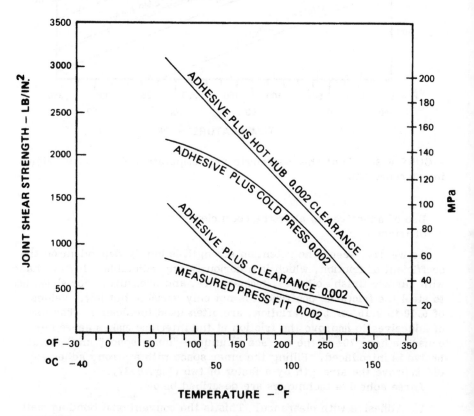

FIGURE 6.37   Comparison of assembly methods—joint strength vs.
temperature.

that does not decrease as quickly as frictional forces do when the hub pressure drops. In assembling the parts, about 50% more assembly force was required than with an ordinary press fit. The adhesive is not a very good lubricant and may be partially curing during assembly because localized pressure and heat is produced. The hub stress produced is the same as that of a press fit without adhesive.

3. Adhesive shrink fit: Ordinary adhesive technique was used with the usual 0.002 in. clearance but this time the hub and shaft were heated to 250°F (120°C), increasing the clearance at temperature to 0.0026 in. (0.066 mm). The parts were quickly coated with Grade U and slipped together. As the assembly cooled, the hub attempted to return to its original 70°F dimension but was prevented from doing so by the rapidly curing adhesive, which acted as a shim. This method bears a resemblance to a shrink fit, but no metal-to-metal contact occurs. The adhesive is maintained in its toughest compressive mode throughout the temperature range. Fill is better because material is being squeezed out during hub contraction and is not scraped off during assembly as in the press fit.

The adhesive shrink fitting technique not only is good for bearings in aluminum but has been used successfully to assemble hardened differential gears onto a carrier for an automobile. The strength obtained was two to six times the strength of the maximum press fit (Sec. 3.4). The hub stresses are less than those produced by a modest press fit, as shown in Fig. 6.38.

Augmenting Press Fits with Space Limitations

Certain types of assembly do not have room for retaining nuts, rings, or heavy press fits. Conveyor rollers, idler and guide pulleys, tension arms, and textile spindles fit into this category. Aircraft ball rod-ends not only have this space limitation but the fit is usually into aluminum or magnesium (Fig. 6.39).

Preventing Damage and Providing
External Alignment

Often the stresses produced by the press fit will damage the shaft or the housing. One of the primary reasons for the adhesive assembly of shafts in small motor armatures is the extra accuracy achieved because the shafts are not bent by the press. Lamination stacks are very difficult to produce with round, straight holes. Shafts pushed into crooked holes will always bend. Before the development of adhesive techniques every rotor-shaft assembly had to go through a straightening operation that cost as much in labor as the assembly did (Fig. 6.40).

The same distortion can ruin the accuracy of bearing support panels in office equipment. Two or three bearings pressed into a large flat

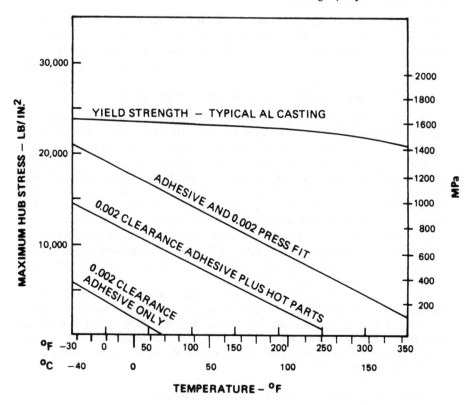

FIGURE 6.38   Comparison of assembly methods—hub stress vs. temperature.

plate or casting can dysfunctionally distort the panel.   In machinist's terms it "warps like a potato chip" (Fig. 6.25).

In addition to maintaining accuracy, an adhesive can be used to obtain fixture accuracy with ordinary parts.   This technique is used to assemble long typewriter platens or rolls.   The rolls are thin tubes with precision bearings cast into the bore with adhesive.   The 1/2 in. (13 mm) shaft can be up to 40 in. (1000 mm) long with bearings spaced intermittently along it for support.   The whole device is slip fitted with adhesive, placed into V blocks to locate centerlines accurately, then induction-heated at the adhesive joints.   The cast-in-place accuracy of these long assemblies is a phenomenal maximum of 0.002 in. (0.051 mm) total indicator runout, or about the equivalent of the shaft's original straightness.   The key to this type of assembly is to have the fixture compensate for part inaccuracies and the adhesive to cast this precision permanently in place.

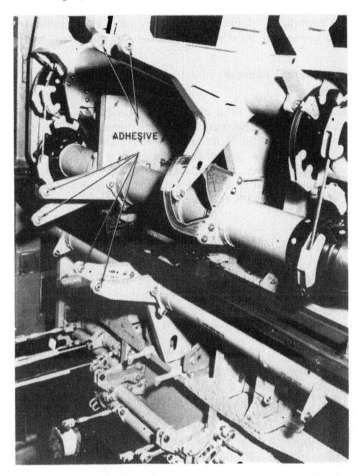

FIGURE 6.39 Ball rod-end for aircraft, bonded and staked.

Providing Fast Repair

The greatest number of adhesive-bearing applications has occurred
in the field where the bearing or the press has failed. Seldom is it
possible to get a replacement for anything other than the bearing; so
damaged shafts and housings must be used until major repairs can be
scheduled. Tales abound about the money and mission saved by replacing a bearing with shims and adhesive . . . like the front-wheel
bearing on a town snow plow, fixed in a blizzard using a torch to warm
the adhesive; or the pulp drier that would not keep a bearing until it
was adhesively mounted; or the railroad track ballast tamper that had
"walking" bearings, fixed on the job. These applications serve to

remind us that all techniques of bearing mounting have their strengths
and limitations and are worth careful consideration by their implementors.

7.3   Mounting of Spherical Plain or
Roller Bearings

Spherically seated bearings will bind up if mounted with a heavy press
fit and are a natural for a friction-augmented light press fit (Fig. 6.41).
The trunion bearings in the USA M-60 tank on which the 155-mm cannon
is pivoted are secured with machinery adhesive (Fig. 6.42). All of the
shock of firing is absorbed through the bearings and transmitted to
the turret, which is secured with bolts locked with Grade K. Any
looseness would wreak havoc with the gunner's aim.

7.4   Limitations

Bearing Float

As mentioned previously, the rear bearing of a spindle usually has
to float to accommodate differential thermal expansion between the spindle and housing. As yet there is no way that machinery adhesives can
improve the fit without increasing the friction at the same time, so they
must be excluded from this area.

FIGURE 6.40 Fractional HP electric motor shaft adhesively assembled
without distortion.

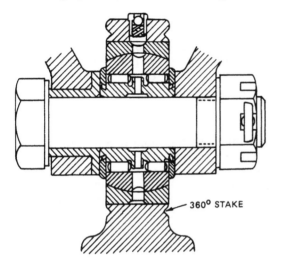

FIGURE 6.41   Cutaway of a spherically mounted roller bearing.

FIGURE 6.42   Trunion bearings for the M-60 tank.

Needle Bearings

Drawn cup needle bearings are not round before mounting. They conform to the housing bore into which they are pressed. Looser fits than those recommended by the manufacturer should not be used, but friction improvement with a machinery adhesive is beneficial.

Excess Material

Care should be exercised in application of adhesive to the outer races. Introduction of the adhesive into the ball raceway, where there is potential for curing even with only line contact, should be avoided. Usually the lubricant in which the bearing is packed will resist entry of excess adhesive. Buna N seals can be attacked by the adhesive and sustained contact should be avoided.

APPENDIX

Design Calculations for Interference Fits

While the design procedures for interference fits are well known, it is worthwhile to review them here to see the effect of dissimilar materials and varying temperatures. (See Fig. 6.43.)

Hub Pressure

The key design parameter for press fits is the radial contact pressure calculated as follows:

$$P = \frac{\delta}{b\ (A + B)} \tag{6.5}$$

where Factor A for Shaft A =

$$\left(\frac{(b^2 + a^2)}{(b^2 - a^2)} - \mu_s\right) \times \frac{1}{E_s}$$

and Factor B for the Hub B =

$$\left(\frac{(c^2 + b^2)}{(c^2 - b^2)} + \mu_h\right) \times \frac{1}{E_h}$$

and P = contact pressure, lb/in^2 (Newton/m^2 or Pa); a = shaft ID, in. (m); b = shaft OD, in. (m); c = hub OD, in. (m); δ = shaft-to-hub interference, in. (m); E = modulus of elasticity, psi (Pa); μ = Poisson's ratio. Subscripts s and h refer to shaft and hub, respectively.

Joint strength, axial capacity and torque capacity can then be calculated as follows [3]:

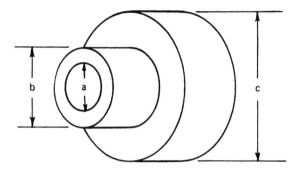

FIGURE 6.43   Shaft-hub press fit.

Joint strength, lb/in^2 (Pa)   (6.6)

   S = Pf

Torque capacity in-lb (Newton-meter)   (6.7)

$$T = \frac{Pf\pi b^2 L}{2} \text{ or } \frac{S\pi b^2 L}{2}$$

Axial capacity, lb (Newton)   (6.8)

   F = PfπbL or SπbL

where f = coefficient of friction; L = engagement length, in. (mm).

The interference causes a tangential stress in the hub which is at a maximum at the hub bore. This stress can be calculated as follows:

Tangential hub stress, lb/in^2 (6.9)

$$S = \frac{P\ (c^2 + b^2)}{c^2 - b^2}$$

Temperature Effects

When materials are assembled that have different coefficients of thermal expansion, consideration must be given to the change in interference when the assembly is heated. If the shaft has a lower coefficient of thermal expansion than the hub, the interference decreases with temperature. If the materials were reversed, the interference would increase when heated. As the temperature changes, the values of radial contact pressure, torque capacity, axial capacity, and hub stress all change due to their dependence on interference.

The change in interference can be calculated as follows:

$$\delta_1 = \delta_0 - b \times (\alpha_h - \alpha_s) \times (T_1 - T_0) \tag{6.10}$$

where δ_0 = initial interference at temperature T_0, in. (m); δ_1 = interference at service temperature T_1, in. (m); b = shaft diameter, in. (m); δ_h = coefficient of thermal expansion of the hub (per °F or °C consistent with $T_1 + T_2$); α_s = coefficient of thermal expansion of the shaft.

REFERENCES

1. J. E. Shigley, *Mechanical Engineering Design*, McGraw-Hill, New York, 1977, Sec. 2–16.

2. *How to Prevent Ball Bearing Failures*, Fafnir Bearing Division of Textron Inc. New Britain, CT, Form 493.

BIBLIOGRAPHY

1. J. L. Sullivan, "Guarding Against Fatigue Failures in Press Fitted Shafts," *Machine Design*, June 9 (1977).

2. E. Kerekes, R. H. Krieble, R. Wittemann, and R. Nystrom, *A Comparison of Holding Power Between Press Fitted and Retaining Compound Bonded Metal Assemblies*, ASME Paper 64-Prod-24, American Society of Mechanical Engineers, New York, 1964.

3. R. Thompson, *Improved Methods for Fastening Steel Parts in Aluminum Housings*, SAE Paper 790503 1979, Society of Automotive Engineers, Warrendale, PA. 1979.

Chapter 7
Sealing: The Art and Science of Preventing Leaks

1. INTRODUCTION

Any discussion of sealing must of necessity consider the converse of sealing: leakage. Leakage exists in all fluid (gas, vapor, or liquid) systems to one degree or another, and the amount of leakage that can be tolerated is a matter of choice. Acceptable leak rates can range from a slight drip to bubble-tight to molecular diffusion through the base materials. Equipment users want trouble-free operation but it is not always practical to specify zero leak rates. Overspecification in this area usually leads to increased costs and sometimes impractical or unwieldy designs. Factors influencing acceptable leak rates are toxicity, product or environmental contamination, combustibility, economics, and personnel considerations. All these factors have received increased emphasis since the oil crisis in the 1970s and '80s. In the heyday of tail fins and V-8 engines, one expected to have oil spots on the floor of the garage where the car was parked. Every car "burned" oil, which had to be replenished every few hundred miles. Now the engine manufacturer's objective is not only to eliminate visible leaks but to have the whole car pass the "shed test," where the car's total emissions must be below stringent levels (Code of Federal Regulations #40, Parts 81—99, July 1979). In the test a car with a hot engine is wrapped in a plastic bag to capture all but exhaust emissions. Escaping hydrocarbons are measured with sensitive instruments that measure to 0.5 parts per million.

Leakage from hydraulic systems alone costs industry more than $250,000 per day. Added to this is the cost of leakage in other fluid systems, such as water, air, steam, and refrigerant.

Leakage can also be hazardous. Discharged fluids leaking onto floors or walkways present a safety hazard to personnel as well as a

fire hazard. From a housekeeping standpoint, leakage is messy and
unsightly and projects an image of sloppy habits. These and other
problems can be eliminated or minimized through proper leakage control.
In addition to savings in material and labor costs, other benefits that
accrue from leakage control include increased production rates, decreased machine down-time, and the prevention of product spoilage.
In general, the safe, continuous, and proper operation of all fluid systems relies on the integrity of the system, and excessive leakage de-
feats this.

2. TRADITIONAL THREADED SYSTEMS

Technological advancements in manufacturing have enabled many fluid
carrying systems to be unitized to eliminate separate components, thus
reducing the number of joints that might leak. This is a good objective
for a designer, but in many cases it is not practical. Accessibility for
repair is still necessary. Many systems, such as water and steam, are
assembled in their place of use with many joints between pumps, valves,
pipe branches, and machinery. These joints are considered to be semi-
permanent and are largely composed of some kind of thread that can
be backed off for repair but is not disturbed during normal use.
    There are four types of threads commonly used for threaded fittings:

1. American standard tapered pipe thread (NPT, National Pipe Thread)
2. Dryseal American standard taper pipe thread (NPTF)
3. American standard straight thread (NC, National Coarse)
4. European series, straight female and tapered male

(See Fig. 7.1.) The most common and least expensive to produce joints
are the tapered thread assemblies. The mating taper threads are
wedged together as they are tightened, with the result that there is
some metal yielding and accommodation of form imperfections during
assembly. Even greater yielding must take place in the mismatch between the taper and straight threads of the European system. The
root and crest of the threads cannot be relied on to seal and therefore
are usually filled with a dope or sealant of some sort. Tapered threads
can be overtightened, creating stresses that can cause immediate failure
of the female part or eventual failure from the fatigue effects of repeated pressure pulses added to a residual stress (Figs. 7.2 and 7.3).
    Straight threads are used where there can be justification of pro-
ducing a separate sealing element outside of the thread. One advantage
of the straight thread is the lower radial force produced during tightening. The sealing element, usually an O ring, is self-energizing from
internal pressure so only good initial contact is necessary to form a
high-pressure seal. There is no benefit from or need for overtightening

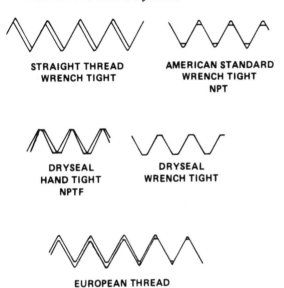

FIGURE 7.1 Straight and tapered thread assemblies.

a straight fitting. As with any thread with axial force on it, thread movement will create a tendency to self-loosen. Since most fittings are connected to a pipe or another fitting this self-loosening is opposed by the other fittings and no harm ensues. *Where the fitting is free-standing, such as a plug or gage fitting, the threads should be locked with something other than torque and friction.* A straight thread will loosen at a torque about 70% of the tightening torque (Chap. 5) and is not as secure as a tapered thread. The tapered thread creates no end thrust during tightening. All torque goes to friction. In tapered threads the tightening and loosening torque are identical. With either thread an auxiliary thread locking material should be used for free standing fittings (Figs. 7.4 and 7.5).

Tube fittings are used to make the transition from a threaded connection to an unthreaded tube. Unlike metal pipe, tubes have relatively

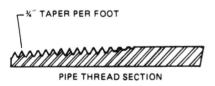

FIGURE 7.2 American standard taper (NPT).

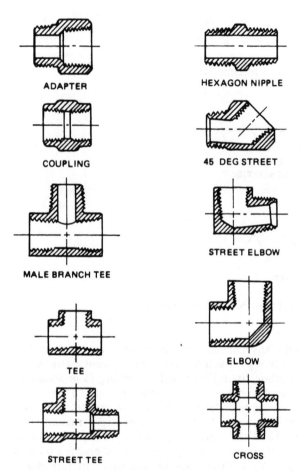

FIGURE 7.3   Typical taper threaded fittings.

thin walls so they can be bent to shape with long-radius bends, which
offer low resistance to flow (Figs. 7.6–7.8).

Tube fittings are subject to the same errors of assembly as any
other fitting. Flares must be accurately formed at time of assembly.
Excessive working or overtightening can cause embrittlement and cracking. Improper lubrication causes pick-up or galling and low seating
force with leaks resulting. Tightening torques are specified but are
usually ignored except in a production situation. Like the assembly
of most fittings, tightening is by feel and large variables in effectiveness result.

Each fitting requires that the proper tightening torque be applied
and that movement of the pipe and fitting in the loosening direction be

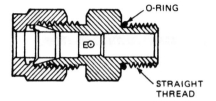

FIGURE 7.4   Nonadjustable straight thread fitting.

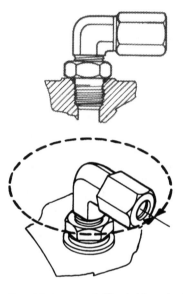

FIGURE 7.5   Adjustable straight thread fitting.

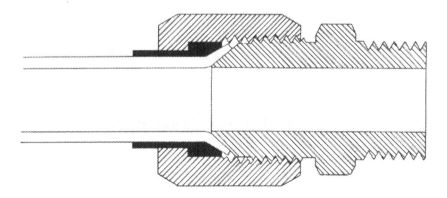

FIGURE 7.6   Typical flared fitting.

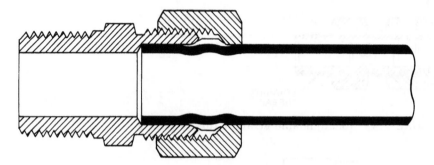

FIGURE 7.7 Flareless compression fitting for copper, aluminum, and plastic tubing.

prevented. Such movements cause either some or all of the contact pressure to be lost, depending on the nature of the joint. For example, Dryseal threads and flared tubing have very little compressibility and small movements in the loosening direction cause immediate loss of the original metal-to-metal contact. The O ring, on the other hand, can accommodate some movement, proportional to the stretch of the elastomeric material used. Vibration and shock loosening of fittings have always been major concerns in controlling leakage. Even though fittings are called static seals, they are subjected to a variety of dynamic forces, which are discussed in the next section on flanged assemblies.

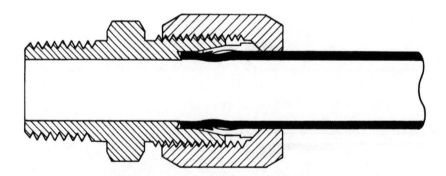

FIGURE 7.8 Flareless tube fitting for high pressure (10,000 lb/in.^{2}, 69 MPa).

3. FOUR SEALING TECHNIQUES

3.1 Trapped Dopes

The oldest and least expensive way to plug the spiral leak path of a
sealing thread is to use drying or nondrying dopes. These are made
of many materials, fillers, and oils, to a thin paste consistency. They
usually have a lubricating and orifice-jamming quality. Their main
problem is that they are inherently weak and extrude readily in large
gaps. If they are made to dry for increased strength they shrink and
eventually leak. Often they are soluble in oils and many chemicals so
they are limited to sealing water and gases.

3.2 Yielding Metal

The sealing interface is limited in area and unlimited in force so yielding
takes place and minute scratches and misalignments are filled by metal
flow. The Dryseal thread is an example of this technique. The roots
and crests of the threads have point interference and flow occurs.
This is effective if the threads are smooth and accurately formed. Production experience shows that metal flow can be relied upon about 98%
of the time. The other 2% is not practical to pursue with decreased
tolerances so a thread sealant is used.

Flares and sleeves used with tubing have the same drawback, being
almost perfect but needing a sealant to make them 100%.

3.3 Trapped Elastomer

The confined O ring is very effective but it too suffers from sloppy
assembly. Excessive tool marks, twisted assembly, and pinched diameter are contributors to leaking failures. The ring is used mostly in
hydraulic systems where the extra cost is more easily absorbed and its
freedom from contamination is desirable.

Two other forms of elastomers are used very effectively in tapered
threads. The best known is Teflon tape. It gives a good initial seal
but can creep away over time. Further, it is often banned in hydraulic
systems because it can be shredded by the threads during assembly.
Being almost indestructible and insoluble, the shreds take up permanent
residence in critical orifices and clearances in the system. However,
because of Teflon's chemical inertness it is the only organic sealant
allowed for sealing liquid or gaseous oxygen.

The other thread sealing elastomer is a preapplied latex that is used
primarily on tapered threads. It is bonded strongly to the male thread
and forms a yieldable interface to accommodate the thread variables and
plug the spiral leak path. It is used extensively for medium pressure
hydraulic, air, and water seals in transportation equipment. Treated
parts can be reused four or five times, about as often as an O ring can.

The problem with elastomers is the question of permanence. They must be selected carefully for compatibility with the fluid being sealed and the environment. They must be carefully confined because they are inherently weak and creepy, but they should not be overconfined because they are incompressible and need space to flow. No thread locking occurs with these materials. Relaxation can occur with creep and shrinkage.

3.4   Curing Resins, Machinery Adhesives

The strength of machinery adhesives is between that of elastomers and that of yielding metal. They are capable of compressive strength equal to that of soft steel and tensile strength 10 times that of elastomers. Clamp loads need be only tight enough to prevent separation in use. Because they develop strength by curing after they are in place they are forgiving of tolerances, tool marks, and slight misalignment. They make permanent tapered-thread joints as effective as O ring-sealed fittings at a fraction of the cost. Used with other methods they can improve effectiveness from the normal 98% to 100% with large savings from eliminating rework of leaky joints. They *must* be used to provide locking of free-standing fittings. Flexure and fatigue failure of flared fittings can be stopped by applying Grade N inside the nut extension.

There are limitations and cautions that should be observed. When paste form W, X, or Z is used care must be taken to push or brush the material into the root of the threads before assembly. These materials are formulated to stay where they are put. After proper assembly and before cure they will resist test pressures of 15 to 50 lb/in.^{2} (100–300 kPa), actual resistance depending on the size of orifice and time of application.

Sensitivity of the system to contamination should be considered. Although Grade W is being used in the most sensitive systems by application to the second and third thread of tapered pipe joints, material can get into the system (as Murphy says, "If something can possibly go wrong, it will"). Grade W is a filled material with lubricating particles that have a size range of 1-40 μ. It would be preferable to use Grade X or other unfilled varieties K, L, R, S, or U where application is critical. The unfilled liquids are usually miscible in hydraulic fluids. They become diluted and ineffective in large amounts of oil. They have a good reputation for being compatible with filters and orifices. They far outperform dopes and the fluoro-silicone tape materials in this respect.

Machinery adhesives are formulated for use on metal parts. If the materials are used on plastics or as a gasket dressing, activator/primer N should be used to activate the surfaces. Plastic pipe threads can be sealed but tests should be conducted for long-term compatibility with stressed fittings. Plastic fittings distort so much during assembly that

drying dopes and solvent cements are used successfully. Higher performance sealers are not necessary.

Limitations of gap curing must be observed. Pipes over 1 in. or
25 mm often exceed 0.03 in. (1 mm) gaps at the crest of the threads,
extending cure time considerably.

4. SEALING APPLICATIONS

Many types of static machinery joints are vulnerable to leakage. They
can be protected with anaerobic sealants in all the applications shown,
subject only to temperature and other environmental conditions. (See
also Figs. 7.9 and 7.10.)

1. Hydraulic joints: Anaerobic sealants seal the ferrule and shock-
   proof the nut threads. They seal and lock tapered and straight
   threads.
2. Tubing joints: Aluminum and steel rolled joints can be sealed without excessive rolling and cold working.
3. Threaded parts: Studs and bolts can be run into reservoirs and
   sealed with anaerobic machinery adhesives Grades K, L, or O.
4. Flanged joints: Gaskets can be replaced or dressed with Grade X.
5. Plugs: Either driven or threaded plugs are sealed.
6. Pipe threads: Heavy-duty service is given with Grade W without
   shredding or dissolving.
7. Shaft assemblies: Bearings and shafts can be sealed against leakage.
8. Shaft seals: Outside diameters can be leakproofed with Grade S.
9. Press fit: Inner space can be sealed at the same time that the
   strength is augmented with any of the liquid grades.
10. Cracked castings: After repair by welding or brazing, the micro-
    porosity can be filled with Grade R.
11. Porosity: Castings and powder metal parts are sealed with anaero-
    bic materials by bulk processing in a vacuum-pressure system.
12. Weld porosity: Microporosity in welds is sealed with Grade R.

5. FLANGED SYSTEMS

5.1 Static Seals

All types of fluid seals perform the same basic function; they seal the
process fluid (gas, liquid, or vapor) and keep it where it belongs.
They accomplish this by forming an impervious barrier against fluid
transfer between two mating surfaces. Traditionally seals are categorized as either static or dynamic. The primary distinction between

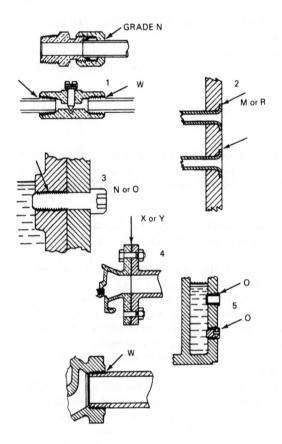

FIGURE 7.9   Applications of machinery adhesives to fluid joints.
(See text for key.)

them involves the degree of movement with respect to the mating surfaces. For example, dynamic seals are used to retain fluids or throttle
leakage between a sliding or rotating part and a stationary one, whereas
static seals prevent fluid loss between two stationary surfaces. This
generic classification of seals is somewhat misleading and implies that
static joints are completely rigid. While there are no gross movements
between the mating parts, micro-movements are present because of
several agents that act alone or in combination with one another. The
four most important agents are temperature changes or differences,
fluid pressure, fluid velocity, and system forces. These agents produce adverse effects on seals in ways that are not always fully ad-
dressed (Fig. 7.11).

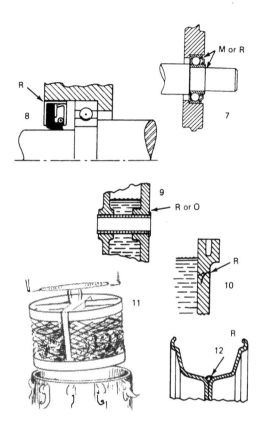

FIGURE 7.9   (Continued)

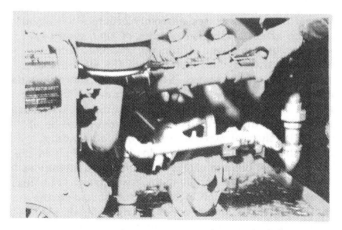

FIGURE 7.10   Fire truck pumping and piping system. "Anaerobic pipe sealant is the first we have ever used that is 100% effective," according to National Foam System Incorporated. "It not only saves us a day or two correcting leaks on new trucks, we also know they will never leak in the field . . . we were able to speed up greatly the assembly of complex pipe systems because the sealant also locks elbows and joints in any position without locking up the threads. This is a major cost-cutting benefit."

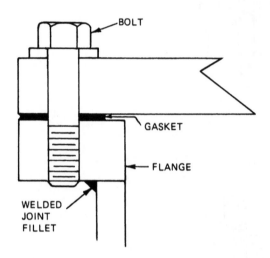

FIGURE 7.11   Typical flanged joint.

5.2   Temperature Effects

Fluctuations in temperature of a gasketed assembly create a variety of
changes, all of which can affect sealing properties of that assembly.
High thermal stresses can be developed within the gasketed joint because of the various sizes, shapes, and materials of the elements involved, i.e., bolts, flanges, and gaskets. Stresses from thermal loads
occur when the components are subjected to temperature gradients or
to uniform temperatures where the components have different coefficients
of thermal expansion. These thermal loads are high enough to be of
engineering concern. The thermal stresses can produce abrading,
crushing, extrusion of the seal, or loss of bolt load. Example 1 in the
appendix shows that a thermal change of 370°F will cause a change in
bolt length by the same amount as a 80,000 lb/in.^{2} prestress (550 MPa).
This means that for each 1°F difference in temperature between the
bolt and the flange, the stress in the bolt changes by 216 psi. Example
2 of differential thermal expansion shows that an aluminum flange bolted
to a steel flange expands radially 0.013 in. more than the steel for
every 10 in. of flange diameter and for a 200°F change in temperature.

5.3 Pressure and Velocity Effects

Fluid pressure and velocity surges can also be present in the system. Surges usually result in shock and vibration. For example, if a valve is suddenly closed or an obstruction blocks the flow of fluid, a pressure wave is generated by the kinetic energy of the fluid. This pressure wave travels at the speed of sound for the fluid through the downstream line until the wave is reflected back to the point of origin. If the system has separate branches, the wave is reflected separately through each branch. This phenomenon is repeated with shock waves overlapping one another until the original kinetic energy is absorbed by friction. In a water system the shock waves are referred to as water hammer because of the noise and shock that usually accompanies them.

5.4 External Forces

Gaskets can be differentiated from sealants by the fact that they are designed to support the clamp load and any external forces on the system, whereas sealants are subjected only to internal pressure and shearing flange movements. Gaskets must have physical pressure against them at all times in order to seal. Sealants maintain their integrity by improving the friction forces on the flanges; all clamping and external loads, except for shear, are borne by metallic contact of the flanges.

External forces stress a gasketed joint in a manner similar to the internal shock waves. An out-of-balance pump can vibrate the connected pipe rather severely, causing premature joint failure. Likewise joints on cars, airplanes, and boats will be subjected to all of the vibration and twisting motions of that vehicle.

In summary, flange systems are not truly static but are a throbbing, squirming, shaking system composed of several elements. These elements are the flanges, bolts, and gasket or seal, and sometimes the bracket attaching the pipe to a shaking vehicle frame. The design material selection, and maintenance of any gasketed joint must include consideration of all the agents acting on the system. Very often the system is a compromise of choices. For instance, the material that will seal rough flanges may not be the material that will withstand hydraulic hammering. Sometimes this compromise means changing some of the elements in order to get an effective joint seal. For ease of analysis, let's first study each of the important elements of a gasketed system individually and not attempt to pull them together until the end of this discussion.

6. FLANGE DESIGN CONSIDERATIONS

Although the gasket material is considered the most important element in the system, it may fail to provide a seal because the other elements were not designed and constructed to make the best use of its properties. The initial or bolting-up pressure must be enough to cause local yielding of the gasket where it is in contact with the asperities left from machining the flange surfaces. This minimum contact pressure, necessary to secure a tight joint even for low values of internal pressure, is called the yield value, y, of the gasket. A few typical values are given in Table 7.1. More values for y can be found along with the gasket factor, m, in the ASME Code for Unfired Pressure Vessels, Section VIII.

Internal fluid pressure in the pipe reduces the gasket contact pressure. Experience has shown that the ratio between the resultant contact pressure and the fluid pressure being sealed should not be less than a certain value if the joint is to remain sealed. This ratio is called

TABLE 7.1   Gasket Factors, m, and Yield Values, y

	Gasket factor, m	Yield value, y	
		lb/in.^{2}	MPa
Gum rubber sheet	0.50	0	0
Hard rubber sheet	1.00	180	1.2
Rubber-covered cloth	0.75	50	0.3
Rubberized asbestos and wire two-ply	2.50	2880	20
Asbestos composition 1/8 in. thick	2.00	1620	11.2
Serrated steel, asbestos-filled	2.75	3650	25
Corrugated copper	3.00	4500	31
Corrugated iron or steel	3.75	5450	38
Solid metal, copper	4.75	13000	90
Solid metal, soft steel	5.50	18000	124

the gasket factor, m, and is primarily but not entirely a gasket characteristic (Fig. 7.12).

The bolting-up and fluid pressures cause distortion in the flanges
and also have an effect on the m ratio. The main effect is a rotation
of the flange around the outer edge of the gasket. The yield value y
is reached before the average pressure would indicate, and further
tightening to achieve the m value may cause gross gasket yielding and
its ultimate destruction by the agents previously discussed (temperature, fluid pressure, fluid velocity, and system vibration).

A great deal of work has been and is being done on flanges and gaskets by a committee of the American Society of Mechanical Engineers
(ASME). The results of their work appear in the ASME boiler code.
Serious high-pressure designs should follow this code to take advantage of their tremendous amount of experience.

The major problem encountered with flanges is distortion. If the
working pressure is too high for the design of the flanges, adverse
flange and bolt distortions will occur. Distortions most commonly found
are: (1) bowing, (2) bolt hole distortions, (3) nonparallelism, and
(4) surface roughness. Each one will be discussed separately.

6.1   Flange Bowing

Fig. 7.13 illustrates the distribution of stress on a flexible flange. The
area where this joint is most likely to leak is right at the center of the
flange (c), where the smallest stress is produced by the bolts and where
the maximum bending occurs from internal pressure. One way to improve this situation is to stiffen the flange to get the result shown in
Fig. 7.14. As obvious as this seems, the economics of producing stiff
flanges often preclude this approach. The bowing of flanges is the
most common reason for leakage.

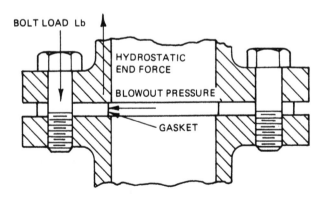

FIGURE 7.12   Forces on a flanged assembly.

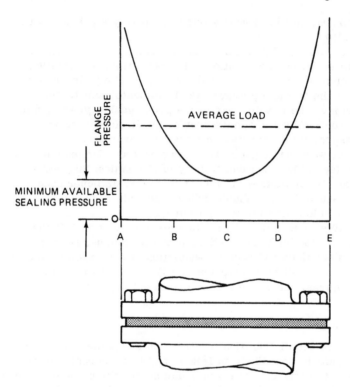

FIGURE 7.13   Stress distribution on a gasket.

6.2   Bolt Hole Distortions

This occurs around the bolt holes in a flange. Fig. 7.15 shows a
stamped sheet-metal flange before bolt loading (bottom) and deformation
after loading (top). High stresses are transferred to the gasket material under the bolt head, causing the gasket to crack, tear, rupture,
or extrude, as shown in Fig. 7.16. Bolt hole distortions do not necessarily lead to leakage but can cause bolt load losses that will result in
fatigue failures if the system is subjected to dynamic forces.

6.3   Nonparallelism

Fig. 7.17 shows flange cocking or nonparallelism. Cocking in itself
does not create serious problems until the pressure in the area of low
gasket compression falls below the minimum sealing stress. This type
of distortion is usually caused by improper machining, improper heat
treating, casting irregularities, or improper tightening sequence of
the bolts (Fig. 7.15).

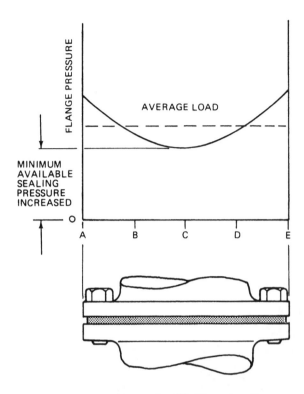

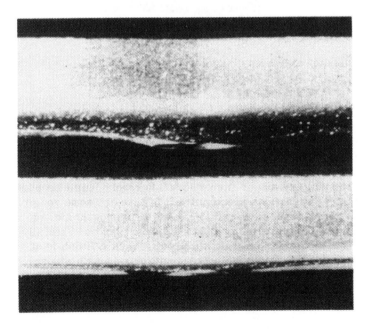

FIGURE 7.14   Effect of stiffening the flange.

FIGURE 7.15   Flange distortion from bolting.

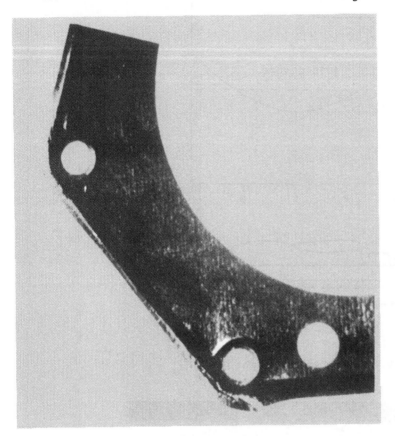

FIGURE 7.16   Gasket extrusion.

6.4   Surface Roughness

The various operations required to fabricate the surface of a flange
face produce distortions or irregularities. A commercial finish is made
by milling machines, grinders, and abraders. The roughest surface
normally found in flange joints is 250 μin. (6.4 μm). On the other
hand, an exceptionally smooth or mirror finish is rarely found because
expensive machining operations are required. There are some rough
surface finishes that are planned and machined as phonographic, or
concentric grooves. They are not random irregularities but carefully
dimensioned in width, height, and depth of cut. Such a finish is almost always found in pipe flange assemblies and is used as a stress
raiser and extrusion preventer. To eliminate leakage from surface
irregularities, the minimum sealing stress must be achieved, thereby
ensuring that the gasket flows into all fange face imperfections.

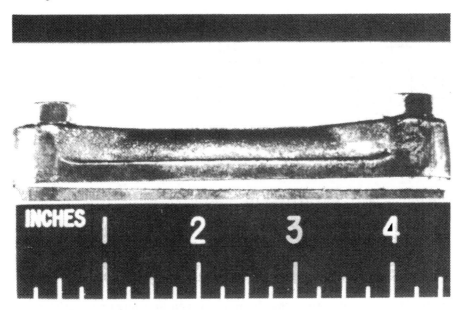

FIGURE 7.17 Nonparallelism.

7. FLANGE BOLTS

Bolts are a critical part of the gasketed joint. The life of the gasketed
joint can be greatly affected by the design of the bolting system. Almost all of the force for producing the minimum sealing stress is generated by the bolts (a small amount may come from gasket adhesion and
gasket swelling from chemical and pressure effects). It is important
that the bolts not only produce the design pressure initially on the
gasket but maintain that pressure throughout the service life of the
assembly. Bolts lose their initial stress or clamp load in a variety of
ways, some of the more important being: (1) gasket relaxation,
(2) bolt hole distortion of the flange, (3) temperature effects, (4) micro-
sliding (unscrewing), and (5) system load changes.

Let's use a spring analogy to help explain the function of the bolts
within a gasketed assembly (Fig. 7.18). The bolts stretch or deform
like a spring when a load is applied. The necessary clamp load cannot
be exerted on the flange and gasket if this axial stretching did not
occur. The amount of clamp load that can be applied depends on the
bolts' strength, the rigidity of the flange, and the compressive strength
of the gasket material. While the assembly is being tightened, the bolt
stretches in tension and the flanges and gasket are compressed. The
goal should be to produce the maximum elongation in the bolt and
the minimum compression of the gasket and flange. Increasing bolt

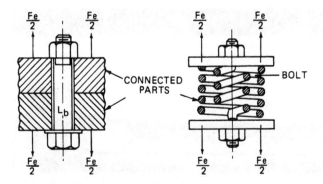

FIGURE 7.18   Spring analogy for modeling a bolt.

elongation helps to maintain the required clamp load. There are two
ways to increase bolt elongation: (1) increase the effective length and
(2) decrease the spring constant by making the bolt smaller in diameter.

Effective bolt length is shown in Fig. 7.19. When the length of bolt
is five times greater than the diameter, it can be elongated sufficiently
to work as a spring between two flanges. The relationship between the
effective bolt length and the clamp load might be understood more clearly with an example. A bolt with an effective length of 1 in. elongates
0.004 in. at 120,000 psi bolt load. If the gasket material relaxed in
compression approximately 0.002 in. over some time period, the original
clamp load would be decreased by 50%. However, if the effective bolt
length were increased to 2 in. under the same load, the bolt would now
elongate 0.008 in. The gasket relaxes about the same amount (the load
has not changed) but this time only 25% of the original clamp load has
been lost. Clearly a 25% load loss is more desirable than 50% because
we are concerned with maintaining minimum sealing pressure.

Several things can be done to increase the effective bolt length:
(1) use a thicker washer under the bolt head, (2) design a boss on
the flange, (3) use a nut instead of a tapped flange, (4) use a thicker

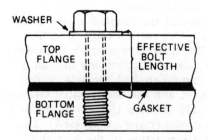

FIGURE 7.19   Effective bolt length.

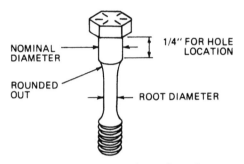

NOMINAL DIAMETER

1/4" FOR HOLE LOCATION

ROUNDED OUT

ROOT DIAMETER

FIGURE 7.20 Lowering of spring constant by reducing the shank diameter of a bolt.

flange, (5) machine down the top portion of the engaged threads to increase the stress in the bolt to the maximum allowable, and (6) decrease the basic bolt size (Fig. 7.20).

Decreasing the size of the bolts may require that a higher grade be used. This is a logical step since the difference in cost between Grades 2, 5, and 8 is minimal compared to the methods of increasing the effective bolt length.

8. GASKETS

The third and final element in the flanged system is the gasket. Its main function is to create a barrier against the transfer of fluid across two mating surfaces which are there to avoid manufacturing costs and provide accessibility. Gaskets are also used for joining dissimilar materials, dampening vibration, and insulating against the transfer of heat.

Materials used for gaskets fall into two broad categories, metallic and nonmetallic. The metallic gaskets can be steel, copper, or a metallic and organic composite. Materials such as asbestos, cork, cellulose, and rubber are nonmetallic gaskets. Sealants are often called gaskets or gasket replacers but are differentiated by the fact that they are not load-bearing. The yield value y of sealants is near zero because they are liquid or pastes as applied (Table 7.1). The ratio m, gasket pressure to sealed pressure, is more a flange requirement than a sealing material requirement. That is, the flange must not bow under pressure and in use enough to exceed the tensile capability of the sealant as an adhesive. Sealants are called formed-in-place materials. These materials can be applied to any shape flange and are generally considered to be pressure-containing, but not load-bearing, because of their fluid consistency upon application.

All gaskets, whether metallic or nonmetallic, must perform four basic functions:

1. Create a seal
2. Maintain the seal
3. Be impervious to fluid flow
4. Be compatible with the environment

Each function has its own relative importance with respect to the integrity of the gasket assembly. A brief discussion of each one will point out its relative importance.

First, a gasket must create a seal between the flanges by conforming to all flange face irregularities, as shown in Fig. 7.21. On smooth or polished flanges a relatively firm material may be used. When surfaces are rough or show excessive tool marks, either a thicker, softer gasket or heavier bolting pressures must be used. The effects on the other elements of changing the gasket thickness will be described shortly.

Second, the gasket must maintain the seal throughout the life expectancy of the joint. Joints are sometimes subjected to considerable movements caused by vibration, mechanical strain, changes in temperature, pressure, and velocity. Despite these movements, the gasket and flange surfaces must remain in intimate contact. The major factor in maintaining contact is the elastic response of not only the gasket, but the bolts and the flanges as well.

NO CONFORMATION (WRONG)

PARTIAL CONFORMATION (WRONG)

TOTAL CONFORMATION (RIGHT)

FIGURE 7.21   Gasket and flange conformation.

Third, the gasket must be impervious to fluid flow, both internal and external. Materials such as cork with rubber or straight rubber normally are impermeable even with small flange loads. Others, such as fiber sheet packings and cork, are impermeable only when compressed sufficiently to close their natural pores.

The fourth requirement is compatibility with the environment. A gasket material must be able to withstand the full range of temperature changes without deteriorating. Flange staining and corrosion, which is generally promoted by vulcanizing agents, accelerators, and moisture present in the gasket material, should not be visible on the flange face. Gaskets must also be relatively inert to the effects of a sealed fluid. Slight swelling is often beneficial, whereas deterioration of the material or contamination of the sealed fluid is not tolerable.

8.1 Physical Properties

Various physical characteristics have been defined to evaluate a gasket's performance properties and to measure its ability to meet the four requirements previously discussed. The most important physical properties of a gasket material are as follows.

Compressibility: This property indicates the degree to which the material is compressed or deformed in thickness by the application of a specific load. This property is expressed in terms of a percentage of the original thickness. Compressibility is used to determine whether a material can compress sufficiently to compensate for surface irregularities or nonparallel conditions in the flange.

Compressive strength: Compressive strength can be described as the maximum compressive stress that a material can withstand without rupture or excessive extrusion. From a functional standpoint, this physical property is important because it is directly related to the ability of a gasket to resist flange loading without breaking down.

Stress relaxation: This is a transient stress strain condition in which the stress decays as the strain or deflection remains constant.

Creep: This is a transient stress strain condition in which the strain increases as the stress remains constant.

Creep relaxation: This is a transient stress strain condition in which the strain increases concurrently with the decay of stress.

Fluid compatibility: Immersion tests provide a simple means of measuring the effect of various liquids on a gasket material. A prepared specimen is immersed in the desired fluid for a specified length of time at a specific temperature. The weight, strength, thickness, or volume changes are the properties most frequently measured.

Flexibility: This property is usually determined by bending a specimen 180° around a mandrel of a specified diameter. As applied to materials in their original condition, flexibility is of value in determining
the handling qualities of the material. It does not directly correlate
with other physical properties.

Heat aging: Heat aging properties are measured by exposing the
material to a specified temperature in a circulating air oven for a specified length of time. Physical characteristics measured after removal
of the test specimen may include flexibility, durometer hardness, and
elongation and are compared with those values for the same materials
prior to testing.

There are many other considerations to be made before a suitable
gasket can be selected, such as its ability to resist tearing, weather,
fire, fungus and, vermin. Most of the physical properties listed above
pertain to cut gaskets; however, formed-in-place materials must also
meet many of these requirements, especially degradation from fluids,
heat, or other environmental constraints.

8.2 Gasket Thickness and Stress Distribution

Perhaps the most important requirement of any gasket material is to
create and maintain the seal. In the section on flanges we defined minimum sealing stress as the minimum pressure the gasket must experience
to close its internal structure to the passage of the sealed fluid and to
conform to the surface irregularities of the flange. As we shall now
see, the thickness of a gasket material has a profound effect on creating
and maintaining the seal.

Figs. 7.13 and 7.14 showed how stiff flanges improved the load distribution on a gasketed joint. To increase the minimum sealing stress
between the bolts, Fig. 7.14 suggested stiffening the flange. The
same effect can be obtained by making the flange bow more, through
the use of a softer or thicker gasket, as shown in Figs. 7.22 and 7.23.

Increasing the thickness of the gasket will enhance its capability to
create the initial seal. Now it would be instructive to see exactly what
effect increasing the gasket thickness has on maintaining the seal. Relaxation is one of the major considerations for maintaining the seal,
primarily because cut gaskets creep or relax in service to some degree.
A relaxation curve of a rubber asbestos gasket is shown in Fig. 7.24.
It shows relaxation of the gasket as measured by torque loss in the
bolt vs. thickness of the gasket. To avoid this effect, use a harder
or thinner material.

Formed-in-place materials should also be considered with respect to
creating and maintaining a seal. Referring again to Figs. 7.22 and 7.23,
flange loading was increased by using a thicker gasket. Now if the
stress distribution were redrawn for a nonadhesive, formed-in-place

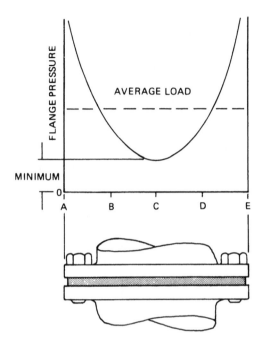

FIGURE 7.22   Stress distribution on a thin hard gasket.

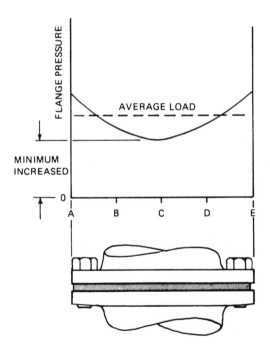

FIGURE 7.23   Stress distribution on a thick soft gasket after compression.

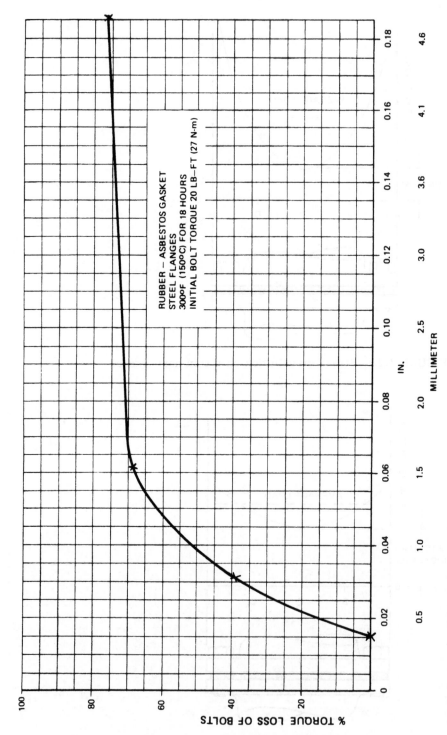

RUBBER – ASBESTOS GASKET
STEEL FLANGES
300ºF (150ºC) FOR 18 HOURS
INITIAL BOLT TORQUE 20 LB–FT (27 N·m)

% TORQUE LOSS OF BOLTS

MILLIMETER

IN.

FIGURE 7.24   Torque loss vs. gasket thickness.

material, the curve would look like that shown in Fig. 7.25. The non-adhesive material will not follow the movements of the flanges. Fig. 7.26 shows what happens if the material is mildly adhesive. A mildly adhesive material will seal flanges that are more flexible, provided that the adhesive is also flexible enough to follow the flange bowing under internal pressure. Usually this adhesive technique cannot be carried to the extreme because flange disassembly becomes very difficult. Fig. 7.27 shows what can be done if the flanges are prestressed in such a manner that internal pressure does not exceed or add to the prestress. In other words, the gasket surface can be prestressed so that the stress distribution is more uniform. This is accomplished by providing prebowed covers or bosses on the mating surfaces (Fig. 7.28).

We have seen that a sealed joint is not a static device. It moves under the influence of temperature, fluid pressure and velocity, and external vibration. If a joint leaks, it may not suffice to change only one element within the system. With any change, one must always keep in mind its effect on the minimum sealing stress.

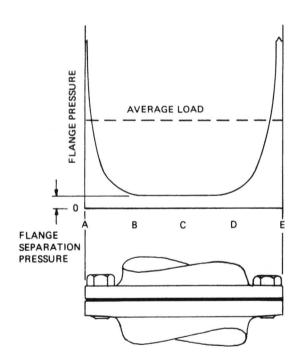

FIGURE 7.25   Nonadhesive sealant.

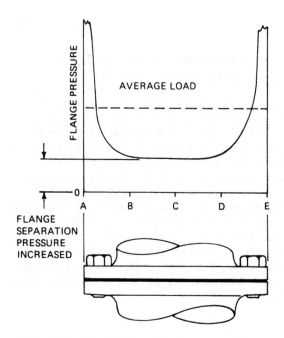

FIGURE 7.26   Mildly adhesive sealant.

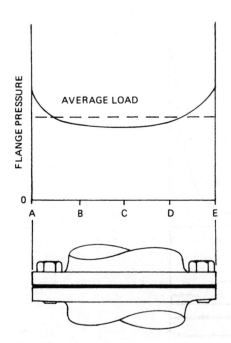

FIGURE 7.27   Adhesive sealant and prestressed cover.

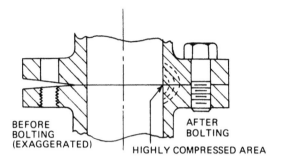

BEFORE
BOLTING
(EXAGGERATED)

AFTER
BOLTING

HIGHLY COMPRESSED AREA

FIGURE 7.28   Flange shaped for prestressing.

9. SYSTEM RELIABILITY

The fatigue or dynamic loading characteristics of an assembly have not
been fully discussed. Their importance to the physical integrity of
the joint cannot be overemphasized because the service life of an assembly is directly related to its ability to absorb and transfer these
dynamic forces.

In most assemblies, the ratio of assembly rigidity to fastener rigidity
is high enough to discount any addition to bolt tension produced by
dynamic forces. In a flexible joint with a soft gasket between bolted
flanges (Fig. 7.29), the joint allows external forces to load the bolt
more than a stiff joint would.

The reason for this may become more obvious by studying the following equation:

$$P = P_i + CF_e \qquad (7.1)$$

where P = final load on the bolt, lb (N); P_i = initial preload or clamping
load developed through tightening, lb (N); F_e = external applied load,
lb (N); and the constant

$$C = \frac{E_b A_b / L_b}{E_b A_b / L_b + E_g A_g / t_g} \qquad (7.2)$$

where E_b = modulus of elasticity of the bolt, lb/in.^{2} (Pa); E_g = modulus
of elasticity of the gasket, lb/in.^{2} (Pa); A_b = effective cross-sectional
area of bolt, in.^{2} (Pa); A_g = loaded area of gasket, in.^{2} (M); L_b = effective length of bolt, in. (M); t_g = gasket thickness, in. (M).

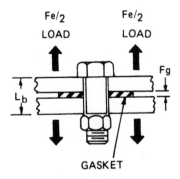

FIGURE 7.29   Flange and bolt schematic.

The value of the constant C falls between 0 and 1. The term $E_g A_g /$ t_g in Eq. 7.2 will be large in comparison to $E_b A_b /L_b$ if the gasket is hard, thin, and large in area, and the constant C approaches zero. When no gasket is used between members in a rigid joint, C = 0. For very soft gaskets, C approaches 1. It is important to remember that Eq. 7.2 is valid only as long as the gasket remains in contact with the flanges. If the bolt stretches to the point where the gasket is no longer in contact, Eq. 7.1 is simply $P = F_e$.

The fatigue strength of a gasketed assembly must be evaluated in two ways: fatigue of the bolt and fatigue of the bolted material. The properly tightened bolt will not fail in fatigue in a rigid joint. Initial bolt tension will stay relatively constant until the external tension load on the joint exceeds the bolt load. If the service load is less than bolt preload, the bolt will experience no failure by fatigue. This is not the case where considerable flexibility is present. Variable stress in screw or bolt fastenings increases with the flexibility of the connected parts. If flexibility is too great, the variable stress present may be high enough to cause eventual fatigue failure of the fastener regardless of the initial bolt preload.

10. DESIGNING WITH FORMED-IN-PLACE SEALANTS

The generic reasons for the use of sealants instead of gaskets are so strong that some of the inherent limitations are worth designing around. Let's look first at the benefits.

1. All creep and relaxation is eliminated.
    a. No retorquing of bolts is necessary. Once the parts are assembled and tested they stay that way. Conventional gaskets gradually relax under stress and must be restressed in the field, the most expensive and least satisfactory place to complete the assembly.
    b. There is positive metal-to-metal transfer of torque and force across the flanges. The sealant must accommodate the flange movements only and not its own deflections from superimposed loads as must a gasket.
    c. There is positive positioning of precision parts. Gear and pump clearances are determined by their position from the machined flange and not from a compressible gasket.
    d. There is no seepage through the sealant.
    e. There are no hazardous materials to be handled such as asbestos-filled cut gaskets. Handling is virtually eliminated in production with automatic equipment.
2. Cost savings.
    a. Inventories are low for formed-in-place sealants. A single cartridge or tube of material will make hundreds of seals. A calculation was made for one high-volume manufacturer of small engines. One year's production requires gaskets that when stacked would form a pile 75,000 ft high. The same capability in sealant is contained in 383 ft of tubes. Various estimates of the cost of inventory have been published. The U.S. Army estimates that it costs $10,000 per year to maintain any separate number in its vast system; for an aircraft manufacture it costs $3,500 and a tractor plant $2,500. To be able to seal 10 different surfaces with one sealant is a saving in inventory before the first application has been made.
    b. Application with screen or transfer is very fast and largely not operator-dependent.
    c. Field repair is easily accomplished with the same low inventory as production.
    d. Intricate parts become simple to cover with the screening or tracing techniques that are commonly available.
    e. Mathematical design analysis of a sealed joint is easier than that of a gasketed joint because there is one less direction of stress. The sealant works in tension only because there is metal-to-metal contact of the flanges.

Now the limitations:

1. The temperatures are limited to the confined (as between flanges) temperature limits of organic materials. This is 300°F (149°C) for all grades in this book except Grade Y, 400°F (204°C), and Grade T, 450°F (232°C). The only organics that will exceed these temperature ratings are the silicone materials, some of which will show fairly good aging resistance up to 600°F (316°C).
2. The flanges must be clamped and bonded so that there is little or no relative movement. The initial bondline is so thin that there is no material to elongate, even with a very flexible material.
3. There is no bending or prestressing of flat flanges so internal pressure may make the unclamped portion of the flange "yawn," putting the sealant in pure tension. The sealant and area involved must be designed to resist yawning, or yawning must be eliminated by shaping one flange. Tension is not the best mode in which to use any sealant.
4. Sealant must be applied on line where a simple, nonautomated approach can become messy, operator-dependent, and more involved than the application of a cut gasket.
   a. The parts must not be grossly contaminated with oil or grease.
   b. Pressure testing should be done within the instant sealing and curing limits of the flanges and sealant.

It is apparent that the substitution of a sealant in a joint designed for a compressible, load-bearing gasket probably will not succeed. This fact has established two rules of thumb for substituting sealants for gaskets.

1. Any gasket over 1/32 in. (1 mm) thick probably will require more than a substitution of material, like the respacing of bolts or stiffening of the flange. One should consider whether the gasket is being used as a spacer or thermal insulator before eliminating it.
2. Using a curing sealant as a gasket dressing for a problem joint overcomes several gasket ailments such as relaxation, creep, exfiltration, and yawning.

11. THREAD AND FLANGE SEALING APPLICATIONS

Outboard motors have almost all the problems of dozens of other applications. The combination of lightweight, thermal loads; salt water; oil; gasoline; hot air; and exhaust in one compact package presents some very interesting sealing problems, many of which have been solved with anaerobic sealants. (See Figs. 7.30–7.32.)

(a)

(b)

FIGURE 7.30   Brass and nylon fittings that feed oil-fuel mixtures into the aluminum cylinder head are locked and sealed with Grade K.   A small·rotary applicator (a) is used to apply conveniently the correct amount to the threads.   Because of the securing ability of Grade K, fittings do not have to be tightened as much as before, eliminating destructive residual stresses.   (Courtesy of Mercury Marine, Division of Brunswick Corp., Fond du Lac, Wisconsin.)

FIGURE 7.31   The pressed diameters of dynamic seals of propeller shafts have proven to be the source of leakage. By coating the outside diameter with Mil-S-22473 Grade A, leakage was eliminated. (Courtesy of Mercury Marine.)

12. POROSITY SEALING

12.1   Powdered Metal

The production of powder metal parts results in porosity, which is inherent to this technology. In some applications the porosity is used either for holding lubricant for better bearing properties or as a precise metallic filter. In many applications the porosity is an undesirable by-product of the process.

The production of powder metal parts consists of four steps:

1. Blending of metal powders with lubricants and binders.
2. Compacting the mixture in a die with high pressure, which causes the particles to bind into a handleable "green" part. After being ejected from the mold, it is placed into trays, which are put on a conveyer belt and moved into an inert-atmosphere oven held, in the case of iron, at over 2000°F (1100°C).

FIGURE 7.32   Rigid tractor transmission flange, which transmits full-wheel torque, sealed with machinery adhesive L.

3. Sintering or fusing of the metal particles occurs at a temperature just below the melting point of the major constituent. Minor constituents, such as copper, may be added to increase strength by melting around iron particles. While sintering is occurring, the lubricants and binders that were necessary to get good compaction for "green" strength are destroyed. They boil off, leaving the

porosity that is powdered metal's inherent characteristic. The
amount of porosity can be controlled between 5 and 50% of the part
by varying the amount of lubricant (usually zinc stearate), particle
size, and compacting pressure. More density than 95% can be produced only by an expensive recompacting process after sintering.
4. Impregnation with a low-melting metal such as copper or an organic
material is usually a final step.

The history of powdered metal processing, which goes as far back
as Osan in 1841, has included impregnation with linseed oil, beeswax,
and varnish, and in more modern times polyester materials, both
styrene- and epoxy-based. In the 1960s anaerobic materials were recognized as ideal impregnants since they cured only in the internal
microporosity, leaving clean the outer surfaces and macroporosity or
holes. Since machinery adhesives Grades R and S are capable of impregnating porous parts, the process is the subject of this section.

12.2   Die and Sand Castings and Welds

The process of pouring molten metal into a mold and allowing it to cool
creates both macro- and microporosity. It is not possible or desirable
to fill macroporosity with an impregnation process. Macroporosity, or
holes over 0.010 in. in diameter (0.25 mm), can interfere with the
strength of the part and should not be routinely plugged with weaker
organics. The two sources of porosity in castings are usually gas absorption and shrinkage during solidification. Usually, gas absorption
by itself does not cause interconnecting porosity; however, with shrinking and crystal formation taking place while gas is being desorbed, interconnecting microporosity is formed. Molten aluminum will pick up
moisture, which dissociates into hydrogen and oxygen. The hydrogen
is only soluble at liquid aluminum temperatures and precipitates out
as it cools.

Cast iron goes through a similar process as it cools with very heavy
precipitation of dissolved carbon into flakes or nodules that are somewhat porous by themselves and when combined with shrink stresses
often produce unacceptable porosity.

Welding of these materials involves melting and cooling similar to the
casting processes. The same type of porosity can occur. Inert gases
are generally used to keep oxides from forming, but microporosity is
a frequent result even in closely controlled automated processes. Typical expectations of leakage by material and process are shown in Table
7.2.

TABLE 7.2 Percentage of Parts Expected to Leak[a]

	% Leakers
Cast iron welds and permanent mold aluminum	2-3
Thin-walled cast iron (less than 1/4 in. [6 mm])	5-10
Aluminum die cast with some machining (intake manifolds and carburetors)	12-15
Heavily machined aluminum die casting (steering gear and compressor housings)	35-90
As-sintered powdered metal	100

[a]95% of all these parts can be salvaged by impregnation.

12.3 Impregnation Processes

One dictionary defines *impregnate* as "to saturate or permeate with another substance." This is the basis of the processes that have been developed for impregnating with one of four materials now commonly used to fill porosity. The four materials are: sodium silicate (or common water glass), heat-cured styrene polyester (300°F, 149°C), heat-accelerated polyester (200°F, 93°C), and the anaerobic polyesters. These materials are used for filling the micropores in iron, steel, copper, aluminum, stainless steel, brass, and fiberglass-reinforced polyester (FRP). The comparison of these processes is shown in Table 7.3.

It is possible with Grades R and S to paint or spray the surface and achieve a shallow penetration that will do the job in many cases. The problem is to effect penetration below the surface. Air is trapped in micropores and must be forced out. The simplest way is to warm the part, expand the air, coat or immerse the part in an anaerobic sealant, and allow the cooling air to pull the material into the pores. This method often works quite well on warm welds, but Grades R and T are very unstable at good air-purging temperatures (300°F, 149°C) and will cure on the surface (Figs. 7.33 and 7.34).

TABLE 7.3 Comparison of Common Impregnation Systems

	Anaerobic polyester[a]	Sodium silicate[b]	Heat-accelerated polyester, low vis.[c]	Styrene polyester, high vis.[d]
Relative cost				
Total	1 (lowest)	2	3	4
Material	$45/gal	$2-2.50	$50/gal	$12.50/gal
Recovery, first try	95%	40-60%	90%	95%
Impregnation method	Wet or dry vac. 40-60°F (4-16°C)	Wet vac. Ambient	Wet or dry vac. 70°F (21°C)	Dry vac. Ambient
Cure method	Warm act. wash 110°F (43°C)	Drying	Hot water 200°F (93°C)	Hot oil 300°F (149°C)
Relative energy use	2	1 (Lowest)	3	4
Temperature resistance	400°F (204°C)	1000°F (538°C)	300°F (149°C)	400°F (204°C)
Solvent resistance	Excellent	Good	Fair	Excellent
Shrinkage	6-12%	40-60%	6-12%	6-12%
Time				
To cure	5 min seal 3-4 hr total	1-30 days	15-20 min	60-90 min
Floor to floor	20 min	30 min	30 min	60-120 min
Viscosity (cP)	10	10	10	250-1000

Maximum pore-filled	0.005 in. (0.1 mm)	0.004 in. (0.1 mm)	0.005 in. (0.1 mm)	0.050 in. (1 mm)
Clean-up	None	Difficult	None	Very difficult
Ventilation needed	No	No	Yes (odor)	Yes (toxic)
Toxicity	Low	Very low	Low	High
Waste disposal	Sewer	Sewer	Some processing	Processing necessary

[a] Centrifuge recovery of material cuts use by 50%. Very clean parts, no bleed. Resin is 100% reactive. Liquids are biodegradable. Cure is insensitive to process time. Process can be done on an assembly line. Cure continues anaerobically to finish at ambient temperatures. Process needs good control to keep material stable. Aerated and cooled resin tank is needed.

[b] Processing equipment is simple to operate.

[c] There is 10–12% bleed-out as the parts are heated to cure temperature. Without heat for proper time the parts will not cure. A resin chiller is needed to keep the impregnation temperature 70°F or lower. Parts emerge hot from the process. The resins are only 90% reactive.

[d] Heat cure causes substantial bleed-out after impregnation. 1 gal of heating oil ($5/gal) is used per 2 gal of resin ($10/gal). Parts are too hot to handle and need degreasing after the cure cycle. Macroporosity can be filled, which may be a problem on intricate parts.

FIGURE 7.33 Automatic transmission housing sealed with a penetrating
spray of anaerobic resin.

FIGURE 7.34   Wheels for tubeless tires sealed at the weld with modified Grade R.

Vacuum Impregnation

The vacuum impregnation process was developed in order to achieve
greater penetration and cleaner parts than are possible with simple
coating.

The process of vacuum impregnating powdered metals and castings
can vary according to the individual requirements of the parts. Similarly, the anaerobic or other materials are fine-tuned to the parts and
the process, and it is recommended that the manufacturer of the impregnation system be consulted. Described below is an acceptable
technique for anaerobic sealing of porous parts that can be handled
in batches (Fig. 7.35).

1.   Clean parts to be processed are loaded into a wire-screen basket.
Parts can be dumped or oriented by hand so that excess material is not
trapped in holes during a subsequent spin-dry operation.

2.   The basket of parts is lowered into the vacuum chamber, completely submerging the parts in the impregnation sealant.

3.   The tank is covered and a minimum 28 in. Hg (711 mm Hg, or
711 torr) vacuum is drawn. The air will bubble from the parts and
from the sealant. The system will stabilize in 1 to 2 minutes.

4.   After the vacuum cycle is complete, ambient pressure is introduced into the tank so that material is pushed into the evacuated pores.
This is an automatically timed cycle and usually takes 8–10 minutes to
achieve maximum penetration.

FIGURE 7.35 Vacuum impregnation processing equipment, including
the vacuum tank with air bubbling and chilling capability, spinner,
activator rinse, and wash.

5. After pressurizing, the basket is raised and spun like a centrifuge for 1 minute to remove excess material from all surfaces and large
holes.

6. The damp parts are immersed in a rinse solution to remove residual surface sealant.

7. A second immersion in a catalyst cures the anaerobic resin at
the pore surface. Resin inside the pores will anaerobically cure as it
is confined away from air and in the presence of metal. The full cure
may take 2 to 3 hours.
Modifications can speed or slow curing or limit the penetration.

Guidelines

Certain guidelines assure maximum effectiveness.

1. The best density range for powdered metal parts is 80 to 90%.
This is 6.2 to 7.0 gm/cc for iron parts. Higher densities limit the
penetration of sealant, and lower densities tend to bleed if heat-cured
materials are used, leaving a rough finish.

2. Cleanliness of parts and the pores relates directly to the quality and performance of the impregnated parts. The optimal time for powdered parts to be impregnated is right after sintering. Tumbling, burnishing, and machining tend to smear the surface, partially blocking the pores. Additionally, these operations are usually accompanied by oils that temporarily block the pores. Conversely, castings should be finish-machined before impregnation since much of the porosity will appear below the cast surface. Care must be taken to remove oils and water.

3. Heat treatment must be done before any impregnation because the impregnants are limited to about 400°F (204°C). Quench oils are particularly hard to remove from pores, and air or water quench is preferred.

4. Powdered metal parts can be coined, sized, or repressed after impregnation. A volume change of 3% is sometimes possible; however, the compressive strength of the parts will be 10 to 15% higher than before impregnation.

Impregnation Applications

Porous parts are vacuum sealed for three application reasons:

Pressure retention
Surface finishing
Machinability

Pressure retention

When parts are designed to hold pressure, interconnected porosity can cause unwanted and unexpected leaks. Powdered metal parts of medium density will usually leak gases and liquids even though about 3% of the total porosity is closed cell or isolated.

Powdered metal parts processed through the vacuum impregnation cycle have been tested to 1000 lb/in.^{2} (7 MPa) in nitrogen and 8000 lb/in.^{2} (55 MPa) in hydraulic fluid.

· Applications for this type of sealing of powdered and cast parts are valves, pumps, meters, compressors, brake pistons, hydraulic parts, engine blocks, heads, and manifolds, air motors, and air cylinder housings. Fig. 7.36 shows an iron-powder based selector control valve cap for a farm tractor hydraulic system. It must withstand a test pressure of 2500 lb/in.^{2} (14 MPa).

Surface finishing

Finishing operations, such as plating or blueing of powdered parts, will leave trapped solutions below the surface if the parts have not been

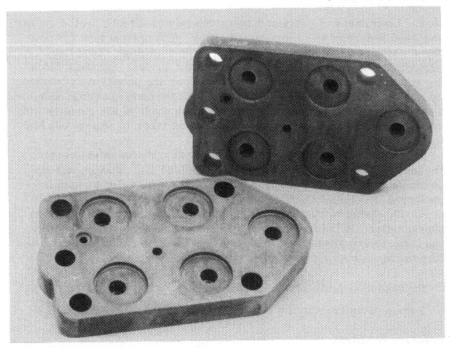

FIGURE 7.36   Iron-powder cap for a selector control valve in a farm
tractor hydraulic system.   It must withstand a test pressure of 2500
lb/in.^{2} (14 MPa).

previously impregnated. The trapped solutions can either corrode the
interior or eat their way through the plating, ruining the finish. Impregnation allows decorative parts to be made by the economical powder
and casting processes. The interior sealed with hard resin prevents
retention of plating solutions and the plating protects the parts from
oxidation. Thr result is extra beauty and durability. Applications for
such surface finished parts are gun and rifle parts (Fig. 7.37), automotive decorative handles, knobs and brackets, and outboard motor
components.

Machinability

On powdered parts particularly, the elimination of air pockets improves the machinability as much as 500%. Porous materials present an
interrupted cut to the cutting tool. At high cutting speed the tool
chatters and tool life is limited. Impregnated parts present a continuous surface and even act as if they are lubricated. Oil should still be
used to cool and lubricate the tool.

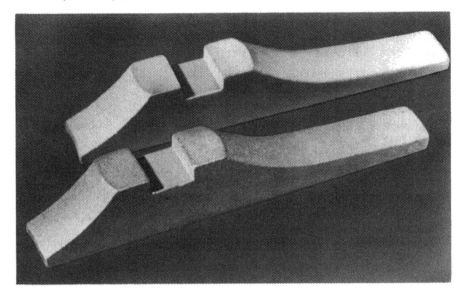

FIGURE 7.37   A gunslight ramp impregnated so that it will accept blueing.

FIGURE 7.38   A micrometer frame impregnated to improve machinability.

Applications for improved machining are those in which the part cannot be completed in a mold or die. These include gun parts, precision tools, and some machinery parts (Fig. 7.38).

APPENDIX

Example 1. Effect of thermal gradients on the bolt: Hooke's Law states that

stress (S) = eE

where e = strain in./in.; E = modulus (for steel = 30×10^6).

The proper preload stress for a Grade 5 bolt is 80,000 lb/in.^{2} thus, the strain or elongation of the bolt is

$$e = \frac{S}{E} = \frac{80,000}{30 \times 10^6} = 0.0027 \text{ in./in. of bolt length}$$

The coefficient of thermal expansion (α) of hard steel is

7.3×10^{-6} in./in./°F

If we let e = change in bolt length for a difference in temperature T, then

e = $\alpha \times$ T

or

$$\Delta T = e = \frac{0.0027 \text{ in./in.}}{7.3 \times 10^{-6} \text{ in./in.}} = 370°F$$

Therefore, for every 1°F change the bolt stress changes 216 lb/in.^{2}

Example 2. Effect of differential thermal expansion, steel flange on aluminum flange:

Coefficient of thermal expansion	in./in./°F
aluminum	= 13.0×10^{-6}
mild steel	= 6.3×10^{-6}
difference	= 6.7×10^{-6}

Assume the parts were assembled at 70°F and working temperature is 270°F or T = 200°F. Then the radial expansion is

$$6.7 \times 10^{-6} \times 200°F \text{ in./in.} = 13.4 \times 10^{-4} \text{ in./in.}$$

or on a 10-in. diameter flange radial expansion will be 134×10^{-4} or 0.0134 in.

BIBLIOGRAPHY

1. "Fastening and Joining," *Machine Design* 14:00 (1967).

2. "Seals," *Machine Design*, June (1969).

3. *Armstrong Gasket Flange Design Manual*, Armstrong Industrial Products Division, Lancaster, Pennsylvania.

4. *Gaskets*, SP-337, Society of Automotive Engineers, May 1965.

5. B. Van den Hoogen and J. H. Dols, Jr., "A New Approach Predicting Tightness of Large Diameter Flanges," *De Ingenieur-Chemische Technick*, May (1970).

6. *Vickers Industrial Hydraulics Manual*, Vickers Inc., Troy, Michigan.

7. *Leakage Control—Hydraulic Lines and Connections*, Mobil Technical Bulletin, Technical Service Division, Mobile Oil Corp., New York.

8. D. E. Czernik, *Recent Developments and New Approaches in Mechanical and Chemical Gasketing*, SAE paper 810367, Society of Automotive Engineers.

9. T. S. Fulda, *Impregnation of Porous Metal Components with Anaerobic Sealants*, SME paper FC77-532, Society of Mechanical Engineers.

Chapter 8
Design Hints

1. INTRODUCTION

Through the years, clever people have developed unique ways to use machinery adhesives. This chapter will attempt to show some of these ways in order to stimulate your thinking about your own design problems. The hints are only hints and are not meant to be complete designs of machine elements. Those we leave to you and your imagination.

2. SIMPLIFYING PART MANUFACTURE

The use of dies and molds to eliminate costly machining operations and to simplify parts manufacture is well known, as are the limitations of these processes. Dies can cut only limited thicknesses. Molds must have draft for removing the part and cannot have under-cuts. By using machinery adhesives to assemble parts, these limitations can be circumvented. You can combine parts made from different processes and materials, such as a steel stamping with a zinc die casting. Powder metal parts can be simplified if they are made in separate molds and combined after sintering with machinery adhesives (Fig. 8.1).

The original sprocket and gear design (Fig. 8.1a) was molded in one piece from powder metal. The clearance-relief in the center was machined after sintering. The machining formed heavy burrs under the gear teeth that were difficult to remove. The compacting and sintering operation resulted in different amounts of shrinkage between the gear and sprocket teeth. Making the parts in separate, simpler molds allowed better shrinkage control and eliminated all machining and deburring by bonding them with machinery adhesives. As a further

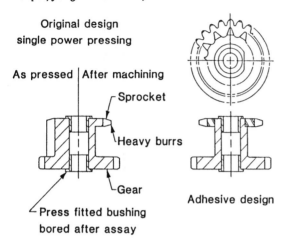

FIGURE 8.1 (Left) original design of a troublesome powder metal gear and sprocket. (Right) redesign using adhesive assembly.

improvement in quality, the bushings were assembled with a slip fit and adhesive while mounted in a centralizing fixture (Fig. 8.1b).

Fig. 8.2 shows how the machining of threads can be eliminated by using a straight bore and an adhesive. A similar process has been used to assemble ceramic knobs with a shaft or a threaded bushing, where a suitable connection could not be produced directly in the ceramic without an adhesive.

Fig. 8.3 shows a sprinkler head that has an adjusting screw locked in place with Grade NN. A lock nut used in this device would have required wrench clearance, which would have complicated the design. The benefits of using a preapplied thread locker are:

Very fine adjustment
Secure locking without disturbing the fine adjustment
Low on-torque for easy assembly
Simple design

Figures 8.4–8.6 show other simplified designs.

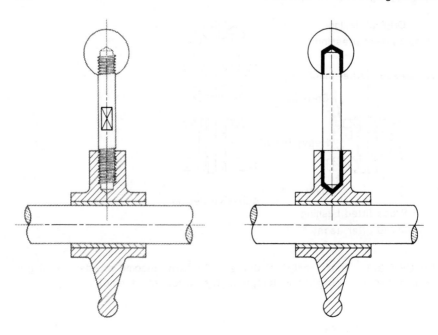

FIGURE 8.2   Elimination of threads to reduce cost.

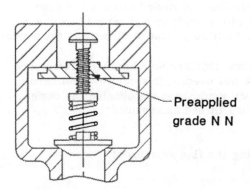

FIGURE 8.3   Pressure relief valve for sprinkler head.

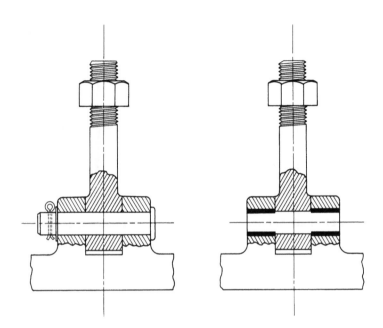

FIGURE 8.4   (Left) high-load clevis.   (Right) simplified design.

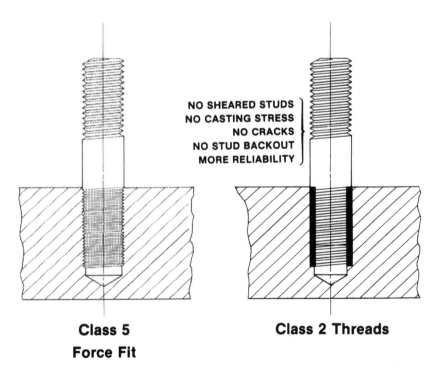

NO SHEARED STUDS
NO CASTING STRESS
NO CRACKS
NO STUD BACKOUT
MORE RELIABILITY

Class 5
Force Fit

Class 2 Threads

FIGURE 8.5   Stud assembly changed from (left) Special Class 5, fine
pitch, to (right) a Standard Class 2 or 3, uniform pitch on each end.

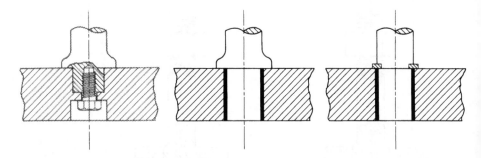

FIGURE 8.6   (Left) shaft anchor. (Center and right) simplified shaft anchor.

3. MAKING USE OF STANDARD MACHINE PARTS

Standard nuts, bolts, bushings, and bearings are available at very low cost. These can be combined into more complex elements with machinery adhesives. Fig. 8.7a shows a bolt and two washers used to make a pivot that can be adjusted to give exactly the right amount of side clearance regardless of the tolerances on the parts. After adjustment the adhesive in the thread is allowed to set before the pivot is used. Fig. 8.7b illustrates how a ball bearing can be incorporated into the design for extreme accuracy and rigidity.

Fig. 8.8 shows how a hardened drill bushing and a soft shaft can take the place of a special hardened shaft. The bushing is used as the seat of a sprag clutch and the wear surface of a shaft seal.

4. CENTERING PARTS

Liquid adhesives cannot provide centering or positioning of parts. Instead of eliminating a press fit, a light press can be maintained for centering. Other ways are shown in Figs. 8.9–8.12. Fig. 8.9 shows how locating ribs can be cast into the tapered bore of a cast or molded part. The taper is designed to allow easy withdrawal of the mold core. Similarly, Fig. 8.10 shows the use of a light knurl to locate a shaft in an armature. The knurl centers the shaft, provides instant fixture, assists in the wetting of the adhesive, enlarges the area wetted, and provides electrical contact.

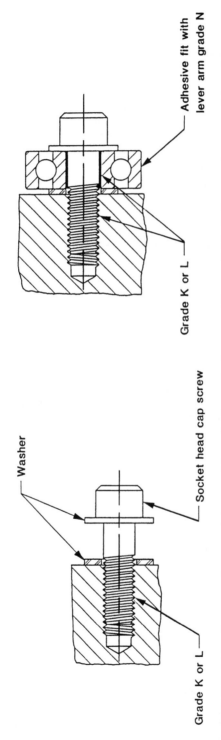

Washer

Grade K or L

Socket head cap screw

Adhesive fit with
lever arm grade N

Grade K or L

FIGURE 8.7   Inexpensive precision machine pivots made from readily available bolts, washers, and bearings.

325

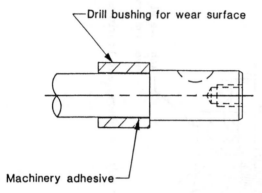

FIGURE 8.8   A hardened drill bushing used as a seat and wear surface for a sprag clutch and shaft seal.

Difficult-to-machine parts can be aligned in fixtures.   The adhesive then acts as a shim as well as a securing means (Fig. 8.11). Fig. 8.12 shows alignment of bushings from a precision ground shaft rather than the widely separated housing bores, which may be quite misaligned.   Machining is simplified, running friction is reduced, and bushing life extended.

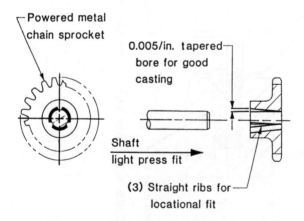

FIGURE 8.9   Straight ribs are cast into the tapered bore of a cast or molded sprocket.

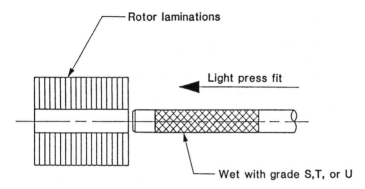

FIGURE 8.10 A light knurl fixtures and aligns a shaft in a motor armature.

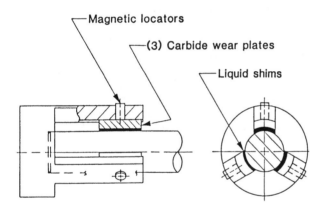

FIGURE 8.11 Carbide wear pads are aligned with an external locating fixture.

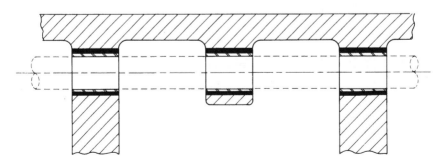

FIGURE 8.12 Bushings are aligned from the shaft.

5. INJECTING ADHESIVE

One of the surest ways to fill between parts is to inject material directly into the center of the joint. Air is purged ahead of the injected material. Fig. 8.13 shows an automated injection process for an air injection valve used on an automobile engine. Parts were dip-activated before assembly for quick cure. The advantages of this method over a press fit or screw thread are:

Ease of automation
Low machining cost, parts used as molded except for injection hole
Excellent retention of short fit regardless of differential thermal
    expansion
Airtight connection

6. SHIMMING AND ELECTRICAL INSULATING

Flat or cylindrical surfaces can be shimmed up to 0.02 in. (0.5 mm) using Grade Z. Surfaces can be leveled with wedges or screws and an indicator or with a fixture as in Fig. 8.11. Material can be pre-applied or injected.

For electrical insulation, adhesive is applied to one surface and dusted with glass beads of appropriate size. The beads provide exact separation of the space, which is filled with the adhesive.

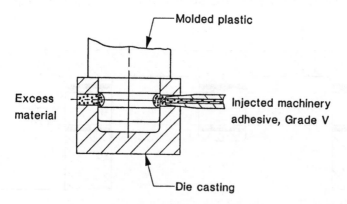

FIGURE 8.13   An air-inlet valve is assembled by adhesive injection.

7. SHORTCUTTING MACHINING

Inexpensive and developmental molds are often made by drilling and reaming locating holes in one slip-fit diameter instead of boring them on a jig borer or grinder in two steps: one for a press fit and the other for a slip fit (Fig. 8.14). Jig boring accuracy can be achieved by reaming the holes for a loose fit and locating mold halves from the cavity. Slipping the pins into place with an adhesive gives excellent alignment. Tolerances between the parts are unnecessary because locating holes and pins are part of a complete assembly.

Fig. 8.15 shows how production parts can be inexpensively doweled together by means of a through-reamed slip-fit hole and an adhesive to secure either one or both ends of the dowel.

Cavities are located from — ⟞ ⟝ Drill and ream through
machined prototype of   for slip fit with dowel
plastic part

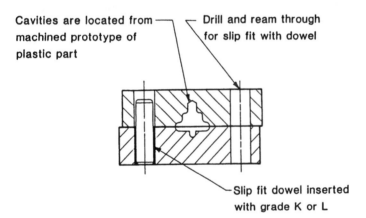

—Slip fit dowel inserted
with grade K or L

FIGURE 8.14   An experimental mold for plastic parts is precisely aligned with bonded dowels.

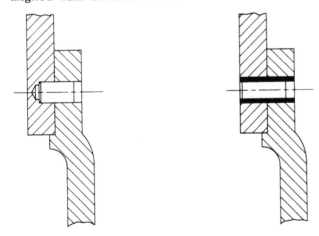

FIGURE 8.15   A slip-fitted dowel is retained and sealed with adhesive.

8. CONCLUSION

By now you should be thinking and planning how you can use space
fillers to bond, seal, and shim those troublesome spaces that every
designer, repairman, and user in years past has taken for granted.
Now you can do something about the limits of friction of a press or
shrink fit. Fluid-tight designs are easy to make without compromising
the design or manufacture with extraordinary attempts to seal cylindrical plugs or threads. Thin castings and powder metal parts become
practical in fluid-tight designs if impregnation is considered as routinely as heat treatment or a coat of paint. A securely locked thread on
a standard bolt can replace the function of special pins, washers, or
nuts. In summary, the use of machinery adhesives in the unavoidable
inner space of fitted parts turns a problem into an asset and improves
the productivity of machinery design, manufacture, and repair.

Glossary

Accelerator: A material that increases the cure speed of an adhesive. It may be either combined with the liquid adhesive or applied to one or both adherends before assembly.

Adherend: A body held to another by an adhesive. (*See also* Substrate.)

Adhesion: The state in which two surfaces are held together by interfacial forces, which may consist of valence forces or interlocking action, or both.

Anaerobic: Pertaining to or caused by the absence of oxygen.

Anaerobic-radical cure: A cure initiated by the loss of oxygen and the presence of free radicals.

Bond strength: The unit load applied by tension, compression, flexure, peel, impact, cleavage, or shear required to break an adhesive assembly, with failure occurring in or near the plane of the bond.

Break-away torque: The maximum torque measured on an unclamped bolt and nut from either friction or an adhesive bond.

Break-loose torque: The maximum torque measured when a clamped bolt is unclamped. It represents a combination of friction and adhesive break.

The author is indebted to the American Society for Testing of Materials for many of the adhesive definitions, which are reprinted with permission from *Standard Terminology of Adhesives*, ASTM D907. Copyright ASTM, 1916 Race Street, Philadelphia, PA 19103.

Break torque: (*See* Break-away torque.)

Catalyst: A substance that markedly speeds up the cure of an adhesive when added in minor quantity as compared to the amounts of the primary reactants. (*See also* Inhibitor.)

Cohesion: The state in which the particles of a single substance are held together by primary or secondary valence forces. As used in the adhesive field, the state in which the particles of the adhesive (or the adherend) are held together.

Creep: The dimensional change with time of a material under load, following the initial instantaneous elastic or rapid deformation. Creep at room temperature is sometimes called cold flow.

Cure: To change the physical properties of an adhesive by chemical reaction, which may be condensation, polymerization, or vulcanization; usually accomplished by the action of heat and catalyst, alone or in combination, with or without pressure.

Diluent: An ingredient added to an adhesive, usually to reduce the concentration of bonding materials. (*See also* Filler.)

Filler: A relatively nonadhesive substance added to an adhesive to improve its working properties, permanence, strength, or other qualities.

Free radical: An atom or compound in which there is an unpaired electron, as $H^{\cdot}$ or $^{\cdot}CH_3$.

Gel: A semisolid system consisting of a network of solid aggregates in which liquid is held.

Glue: Originally, a hard gelatin obtained from animal hides, tendons, cartilage, bones, etc. Also, an adhesive prepared from this sub-. stance by heating with water. Through general use the term is now synonymous with the term *adhesive*.

Glue line (bondline): The layer of adhesive that attaches two adherends.

Hot-melt adhesive: An adhesive that is applied in a molten state and forms a bond by cooling to a solid state.

Inhibitor: A substance that slows down chemical reaction. Inhibitors are sometimes used in certain types of adhesives to prolong storage or working life.

Lap joint: A joint made by placing one adherend partly over another and bonding the overlapped portions.

m: Value for gasket material representing the ratio of gasket contact pressure to pressure of sealed fluid.

Monomer: A relatively simple compound that can react to form a polymer. (*See also* Polymer.)

Newtonian fluid: Any fluid exhibiting a linear relationship between the applied shear stress and the rate of deformation.

Plasticizer: A material incorporated in an adhesive to increase its flexibility, workability, or distensibility. The addition of the plasticizer may cause a reduction in melt viscosity, lower the temperature of the second-order transition, or lower the elastic modulus of the solidified adhesive.

Polymer: A compound formed by the reaction of simple molecules having functional groups that combine to form high-molecular weights under suitable conditions. Polymers may be formed by polymerization (addition polymer) or polycondensation (condensation polymer). When two or more monomers are involved, the product is called a copolymer.

Polymerization: A chemical reaction in which the molecules of a monomer are linked to form large molecules whose molecular weight is a multiple of that of the original substance. When two or more monomers are involved, the process is called copolymerization or heteropolymerization.

Prevail torque or prevailing torque: A fastener torque measured after the bond is broken or while friction alone is resisting turning. Per Mil-S-22473 and 46163, the average of four readings at 90, 180, 270, and 360° from the break. Commercial testing traditionally, and more conveniently, approximates (±10%) this figure with one reading at 180° from the break.

Primer: A coating applied to a surface, prior to the application of an adhesive, to improve the performance of the bond. Sometimes imprecisely used synonymously for accelerator.

Proof load: That bolt load which is guaranteed by the manufacturer to be below the yield point; the safe load for a reusable fastener.

Qualification test: A series of tests conducted by the procuring activity, or an agent thereof, to determine conformance of materials, or materials system, to the requirements of a specification which normally results in a qualified-products list for the specification.

Resin: A solid, semisolid, or pseudosolid organic material that has an indefinite and often high molecular weight, exhibits a tendency to flow when subjected to stress, usually has a softening or melting range, and usually fractures conchoidally (a shell-like fracture).

Rheology: The study of the deformation and flow of matter as indicated by viscosity and thixotropy.

Set: To convert an adhesive into a fixed or hardened state by chemical or physical action, such as condensation, polymerization, oxidation, gelation, hydration, or evaporation of volatile constituents. (*See also* Cure.)

Storage life: The period of time during which a packaged adhesive can be stored under specified temperature conditions and remain suitable for use. Sometimes called *shelf life*.

Structural adhesive: A bonding agent used for transferring functional loads between adherends exposed to service environments typical for the structure involved.

Substrate: The material on which an adhesive or coating is spread for any purpose, such as bonding or coating. A broader term than *adherend*.

Le Système International d'Unites (SI): International system of metric units which forms the basis of the simplified, coherent, metric dimensions used in most countries of the world. (A full treatment of the system is given in ASTM *Standard for Metric Practice* E-380.)

Thermoplastic: Capable of being repeatedly softened by heat and hardened by cooling; a material that will repeatedly soften when heated and harden when cooled.

Thermoset: A material that will undergo or has undergone a chemical reaction leading to a relatively infusible state. *Thermosetting*, having the property of undergoing a chemical reaction iniatiated by heat, catalysts, ultraviolet light, anaerobic-radical, etc., leading to a relatively infusible state.

Thixotropy: A property of an adhesive to thin upon isothermal agitation and to thicken upon subsequent rest. A non-Newtonian behavior where viscosity diminishes as the shear rate is increased. (A fluid is said to exhibit Newtonian behavior when the rate of shear is proportional to the shear stress.)

Viscosity (absolute or dynamic): Resistance to flow as measured by the ratio of the shear stress existing between laminae of a moving fluid and the rate of shear between these laminae. The metric (cgs) unit is the poise having dimensions of gm/cm/sec. The customary unit is the centipoise, having the value of poise $\times 10^{-2}$. The dynamic viscosity of water at 20°C is 1.0 centipoise. In SI, dimensions are $N \times m^{-2} \times s$ or Pa·s, which is equal to cP $\times 10^{-3}$.

y: Value for gasket material. Maximum gasket contact pressure needed to effect a seal.

Index

Milton Keynes UK
Ingram Content Group UK Ltd.
UKHW020017071024
449327UK00032B/2827